Objectives of Biochemistry and Forestry

Objectives of Biochemistry and Forestry

Lakshmi Prasanna Jakka

RANDOM

RANDOM PUBLICATIONS

NEW DELHI (INDIA)

Objectives of Biochemistry and Forestry

ISBN 978-93-5111-799-5

Published in 2016 in India by

RANDOM PUBLICATIONS

4376-A/4B, Gali Murari Lal, Ansari Road
New Delhi-110 002
Phone : +9111-43580356, 011-23289044, 011-43142548
e-mail: sales@randompublications.com,
info@randompublications.com, randomexports@gmail.com

Reprinted 2023

Type Setting by : Friends Media, Delhi-110089
Digitally Printed at : Replika Press Pvt. Ltd.

Preface

Biochemistry, sometimes called biological chemistry, is the study of chemical processes within and relating to living organisms. By controlling information flow through biochemical signaling and the flow of chemical energy through metabolism, biochemical processes give rise to the complexity of life. Over the last decades of the 20th century, biochemistry become so successful at explaining living processes that now almost all areas of the life sciences from botany to medicine to genetics are engaged in biochemical research. Today, the main focus of pure biochemistry is in understanding how biological molecules give rise to the processes that occur within living cells, which in turn relates greatly to the study and understanding of whole organisms.

A forest is a large area of land covered with trees or other woody vegetation. Hundreds of more precise definitions of forest are used throughout the world, incorporating factors such as tree density, tree height, land use, legal standing and ecological function. According to the widely-used United Nations Food and Agriculture Organization definition, forests covered an area of four billion hectares (15 million square miles) or approximately 30 percent of the world's land area in 2006. Forests are the dominant terrestrial ecosystem of Earth, and are distributed across the globe. Forests account for 75% of the gross primary productivity of the Earth's biosphere, and contain 80% of the Earth's plant biomass. Forests at different latitudes form distinctly different ecozones: boreal forests near the poles tend to consist of evergreens, while tropical forests near the equator tend to be distinct from the temperate forests at mid-latitude. The amount of precipitation and the elevation of the forest also affects forest composition.

– Author

Contents

1

Forestry : An Introduction

UTILITY OF FORESTS

In the economy of man and of nature forests are of direct and indirect value, the former chiefly through the produce which they yield and the latter through the influence which they exercise upon climate, the regulation of moisture, the stability of the soil, the healthiness and beauty of a country and allied subjects.

The *indirect* utility will be dealt with first. A piece of land bare of vegetation is, throughout the year, exposed to the full effect of sun and air currents and the climatic conditions which are produced by these agencies. If, on the other hand, a piece of land is covered with a growth of plants and especially with a dense crop of forest vegetation, it enjoys the benefit of certain agencies which modify the effect of sun and wind on the soil and the adjoining layers of air. These modifying agencies are as follows:

- The crowns of the trees intercept the rays of the sun and the falling rain; they obstruct the movement of air currents and reduce radiation at night.
- The leaves, flowers and fruits, augmented by certain plants which grow in the shade of the trees, form a layer of mould, or humus, which protects the soil against rapid changes of temperature and greatly influences the movement of water in it.
- The roots of the trees penetrate into the soil in all directions and bind it together. The effects of these agencies have been observed from ancient times and widely differing views have been taken of them.

Of late years, however, more careful observations have been made at so-called parallel stations, that is to say, one station in the middle of a forest and another outside at some distance from its edge, but otherwise exposed to the same general conditions. In this way, the following results have been obtained:

- Forests reduce the temperature of the air and soil to a moderate extent and render the climate more equable.

- They increase the relative humidity of the air and reduce evaporation.
- They tend to increase the precipitation of moisture. As regards the actual rainfall, their effect in low lands is *nil* or very small; in hilly countries it is probably greater, but definite results have not yet been obtained owing to the difficulty of separating the effect of forests from that of other factors.
- They help to regulate the water supply, produce a more sustained feeding of springs, tend to reduce violent floods and render the flow of water in rivers more continuous.
- They assist in preventing denudation, erosion, landslips, avalanches, the silting up of rivers" and low lands and the formation of sand dunes.
- They reduce the velocity of air-currents, protect adjoining fields against cold or dry winds and afford shelter to cattle, game and useful birds.
- They may, under certain conditions, improve the healthiness of a country and help in its defence.
- They increase the beauty of a country and produce a healthy aesthetic influence upon the people.

The *direct* utility of forests is chiefly due to their produce, the capital which they represent and the work which they provide. The principal produce of forests consists of timber and firewood. Both are necessaries for the daily life of the people. Apart from a limited number of broad-leaved species, the conifers have become the most important timber trees in the economy of man. They are found in greatest quantities in the countries around the Baltic and in North America. In modern times iron and other materials have, to a considerable extent, replaced timber, while coal, lignite and peat compete with firewood; nevertheless wood is still indispensable and likely to remain so.

This is borne out by the statistics of the most civilized nations. Whereas the population of Great Britain and Ireland, during the period 1880-1900, increased by about 20%, the imports of timber, during the same period, increased by 45%; in other words, every head of population in 1900 used more timber than twenty years earlier. Germany produced in 1880 about as much timber as she required; in 1899 she imported 4,600,000 tons, valued at £14,000,000 and her imports are rapidly increasing, although the yield capacity of her own forests is much higher now than it was formerly. Wood is now used for many purposes which formerly were not thought of. The manufacture of the wood pulp annually imported into Britain consumes at least 2,000,000 tons of timber.

A fabric closely resembling silk is now made of spruce wood. The variety of other, or minor, produce yielded by forests is very great and much of it is essential for the well-being of the people and for various industries. The yield of fodder is of the utmost importance in countries subject to periodic droughts;

in many places field crops could not be grown successfully without the leaf-mould and brushwood taken from the forests. As regards industries, attention need only be drawn to such articles as commercial fibre, tanning materials, dye-stuffs, lac, turpentine, resin, rubber, guttapercha, &c. Great Britain and Ireland alone import every year such materials to the value of £12,000,000, half of this being represented by rubber. The *capital* employed in forests consists chiefly of the value of the soil and growing stock of timber.

The latter is, ordinarily, of much greater value than the former wherever a sustained annual yield of timber is expected from a forest. In the case of a Scotch pine forest, for instance, the value of the growing stock is, under the above-mentioned condition, from three to five times that of the soil. The rate of interest yielded by capital invested in forests differs, of course, considerably according to circumstances, but on the whole it may, under proper management, be placed equal to that yielded by agricultural land; it is lower than the agricultural rate on the better classes of land, but higher on the inferior classes. Hence the latter are specially indicated for the forest industry and the former for the production of agricultural crops. Forests require *labour* in a great variety of ways, such as:

- General administration, formation, tending and harvesting;
- Transport of produce;
- Industries which depend on forests for their prime material. The labour indicated under the first head differs considerably according to circumstances, but its amount is smaller than that required if the land is used for agriculture. Hence forests provide additional labour only if they are established on surplus lands. Owing to the bulky nature of forest produce its transport forms a business of considerable magnitude, the amount of labour being perhaps equal to half that employed under the first head. The greatest amount of labour is, however, required in the working up of the raw material yielded by forests. In this respect attention may be drawn to the chair industry in and around High Wycombe in Buckinghamshire, where more than 20,000 workmen are employed in converting the beech, grown on the adjoining chalk hills, into chairs and tools of many patterns.
  Complete statistics for Great Britain are not available under this head, but it may be mentioned that in Germany the people employed in the forests amount to 2.3% of the total population; those employed on transport of forest produce. 1%; labourers employed on the various wood industries, 8.6%; or a total of 12%. An important feature of the work connected with forests and their produce is that a great part of it can be made to fit in with the requirements of agriculture; that is to say, it can be done at seasons when field crops do not require attention. Thus the rural labourers or small farmers can earn some

money at times when they have nothing else to do and when they would probably sit idle if no forest work were obtainable.

Whether, or how far, the utility of forests is brought out in a particular country depends on its special conditions, such as:

- The position of a country, its communications and the control which it exercises over other countries, such as colonies;
- The quantity and quality of substitutes for forest produce available in the country;
- The value of land and labour and the returns which land yields if used for other purposes;
- The density of population;
- The amount of capital available for investment;
- The climate and configuration, especially the geographical position, whether inland or on the border of the sea, &c. No general rule can be laid down, showing whether forests are required in a country, or, if so, to what extent; that question must be answered according to the special circumstances of each case.

These data exhibit considerable differences, since the percentage of the forest area varies from 3.5 to 50 and the area per head of population from 07 to 9.5 acres. Russia, Sweden and Norway may as yet have more forest than they require for their own population. On the other hand, Great Britain and Ireland, Germany, Denmark, Portugal, Holland and even Belgium, France and Italy have not a sufficient forest area to meet their own requirements; at the same time, they are all sea-bound countries and importation is easy, while most of them are under the influence of moist sea winds, which reduces to a subordinate position the importance of forests for climatic reasons.

Intimately connected with the area of forests in a country is the state of ownership - whether they belong to the state, corporations or to private persons. Where, apart from the financial aspect and the supply of work, forests are not required for the sake of their indirect effects and where importation from other countries is easy and assured, the government of the country need not, as a rule, trouble itself to maintain or acquire forests. Where the reverse conditions exist and especially where the cost of transport over long distances becomes prohibitive, a wise administration will take measures to assure the maintenance of a suitable proportion of the country under forest. This can be done either by maintaining or constituting a suitable area of state forests, or by exercising a certain amount of control over corporation and even private forests. Such measures are more called for in continental countries than in those which are sea-bound, as is proved by the above statistics.

THE DEVELOPMENT OF FORESTRY

A feature of fundamental importance, which distinguishes forests from some other resources, is that they are renewable. This means that the mature forest, as it is cut, may be replaced by a new crop of trees; thus a continuous yield of wood is provided from the same area of land. The application of this principle is the primary basis of forestry practice, which may be defined briefly as the growing and using of successive crops of timber.

As the practice of forestry develops, the forests often come to be looked upon less as a natural resource and more as a crop that is dependent in large measure on man's knowledge and skill. However, forestry is not concerned exclusively with growing timber as a crop; forests contribute in many other ways to the well-being of mankind and proper forestry practices assure the maintenance and enlargement of these benefits. Man is not a conservationist by nature: history shows that as long as the existing forest appears to be adequate to the demand, it is subject to exploitation for immediate revenue and there is little concern for its future productivity.

In other words, forest conservation, involving some investment of time and money, only begins when it is apparent that current practices are endangering the resource with critical effects upon the supply of forest products and services. Down through the centuries, as man has increased in numbers and spread over the face of the earth, he has caused and is still causing widespread destruction of the forest cover, partly of necessity to provide land for agriculture, settlement and industrial development, partly through ignorance and greed and partly in the belief that the forest resource is inexhaustible. This latter viewpoint is to some extent justified in the early development of countries and regions that have extensive forest wealth and relatively small populations. Here the forest is often looked upon as a hindrance to settlement and agricultural development, with the result that areas cleared may exceed requirements and include lands quite unsuitable for the purpose in mind.

Man has in fact retained under intensive agriculture only a small portion of the area that was originally cleared for this purpose. It is estimated that, as a result of his activities, man has destroyed about two thirds of the world's original forest cover; the remaining forests containing merchantable timber today occupy less than one fifth of the world's land area. In some parts of the world forest destruction has been unchecked, despite the adverse effect that this has had on the general economy and on the living standards of the people. Asia, in particular, has suffered from the effects of forest clearance and forest destruction, with widespread soil deterioration, inadequate supplies of wood to meet basic domestic needs and low living standards for a large portion of the population. Elsewhere, as the danger became apparent, methods of managing the forests to ensure their continued productiveness were evolved and thus the science of forestry came into being. However, the philosophy of conservation has developed slowly and forestry has invariably been the child of necessity: only after periods of forest destruction and land clearance has the need for husbandry and planned forest economy been realized.

TRANSITION FROM EXPLOITATION TO MANAGEMENT

The transition from exploitation to management takes time and requires the solution of many problems, technical, economic and administrative. In the beginning it may be characterized by a strong reaction against spoliation aimed at preserving the forest intact — "locking it up" — for posterity. In general, however, this is the antithesis of sound management of a renewable resource, which for its fullest development requires periodic harvesting and regeneration; an exception would be the preservation of representative samples of natural forest for scientific study and as "wilderness" areas. Modern forestry originated in Germany and France and much of its development during the past two centuries has taken place in northwestern Europe. Attention to forest conservation in these countries came early owing to an increasing demand for timber and firewood and the difficulty of securing wood cheaply from other countries.

These conditions and the serious effects of deforestation on stream flow, particularly in montane regions, made evident an urgent need for forest management. During the nineteenth century expanding populations and the iron and coal economy of the industrial revolution created new demands for wood and gave added impetus to the development of forestry.

Slowly and usually only as the need became urgent, other countries have followed the lead of western Europe and developed forestry practices to meet their own requirements. Such developments usually begin with restriction on the use of the forest, as in the setting aside of timber reserves and with protection of the forest against its more serious enemies, particularly fire. These steps are followed by efforts to secure new growth after cutting, by encouraging natural reproduction and by planting: with this the practice of silviculture begins. Finally, the accessible, productive forests are managed under cutting regulations and suitable silvicultural techniques to ensure a continuous yield of timber in perpetuity. Such forests are said to be managed on a "sustained-yield" basis.

While the natural forest usually provides the starting point for a forestry programme, it may not represent the optimum conditions that an area is capable of and its perpetuation is not always looked upon as the ideal objective of forest management. The aim, rather, may be to improve the natural forest as far as possible within the limits set by the local environment, silvicultural knowledge and economic considerations and in view of the purpose for which the forest is to be used. Attempts to go beyond these limits and completely change the natural forest pattern have sometimes failed. A well-known instance was the century-long attempt to convert the lowland hardwood forests of Saxony to pure Norway spruce.

Despite early rapid growth, the spruce was often badly damaged by snow and frost as well as by fungal and insect attacks. By the end of the third crop of spruce, conditions had so deteriorated that hardwood species had to be reintroduced and it was generally conceded that the plan had been unsuccessful. On the other hand, there are many examples of the successful introduction of types and species of trees different from those in the natural forest. The success of these operations depends largely on careful study of the ecological factors involved. The aim of forestry may vary from one country to another and even between regions within the same country, depending upon local conditions. Usually the primary purpose is to grow sufficient wood to meet a country's requirements — first for domestic consumption, then for export to foreign markets. Products other than wood, usually referred to as minor forest products, may sometimes be an objective of management. These vary in importance and kind from one country to another, *e.g.*, turpentine, resin, oils, fruits, maple syrup, bark for tanning, etc. The southeastern United States and the Landes in Gascony, France, are two regions where the production of resin and turpentine from species of pine is a major industry carried on in association with timber production.

While the production of wood and other commodities is often primary, in its broader aspects forestry is concerned with the interaction between the forest and its environment, as, for example, in the effects of the forest on the stabilization of soil, on the maintenance and restoration of soil fertility, on the regulation of stream flow and on the provision of a suitable environment for valuable fish and wild life. The protective influences of the forest are many and far-reaching and under certain conditions may be more important than the supplying of wood.

This is particularly true in some parts of the tropics and subtropics, in crucial drainage areas and in regions of light, sandy soil unsuitable for cultivation, where the removal of the forest cover often results in near-desert conditions. Forestry may also aim at the full development of the recreational value of an area; but even where recreation or protection are the main objectives, controlled cutting of wood can usually be carried on advantageously, if the cutting operations do not interfere with the primary purpose of management.

The term "multiple use" is widely used in referring to the management of a forest area for two or more purposes, one of which may be dominant, but not exclusive.

Bases of Forestry

Forestry may be defined as the scientific management of forests for the continuous production of goods and services. While its practice involves many problems of an economic and administrative nature, the solution of which is essential for success, its proper development must in the first place be based on a sound knowledge of the forests and of the laws that govern their development and on an understanding of the characteristics of the individual species that make up the forests and of the interrelationship of the forests with their environment.

The forests of the world include a great many species of trees, not necessarily having any close botanical relationship, with varying adaptations to climate, soil and topography and more or less distinct reproductive characteristics, forms and rates of growth, sizes and ages at maturity, light requirements and abilities to resist damage by insects and other destructive agencies. It is with an intimate knowledge of these features of the trees and tree communities that sound forestry practice must begin.

Supply of Timber : Imports and Exports

The net imports and exports of European countries (average data, calculated from the returns of recent years). The only timber-exporting countries of Europe are Russia, Sweden, Norway, Austria-Hungary and Rumania; all the others either have only enough for their own consumption, or import timber. Great Britain and Ireland import now upwards of 10,000,000 tons a year, Germany about 4,600,000 tons and Belgium about 1,300,000 tons.

Holland, France, Portugal, Spain and Italy are all importing countries, as also are Asia Minor, Egypt and Algeria. The west coast of Africa exports hardwoods and imports coniferous timber. The Cape and Natal import considerable quantities of pine and fir wood. Australasia Net Imports and Exports of European Countries. These net imports are received from non-European countries. They consist chiefly of valuable hardwoods, like teak, mahogany, eucalypts and others. Exports hardwoods and some Kauri pine from New Zealand, but imports larger quantities of light pine and fir timber.

British India and Siam export teak and small quantities of fancy woods. The West Indies and South America export hardwoods and import pine and fir wood. The United States of America will not much longer be a genuine exporting country, since they import already almost as much timber from Canada as they export. Canada exports considerable quantities of timber. The Dominion has still a forest area of 1,250,000 sq. m., equal to 38% of the total area and giving 165 acres of forest for every inhabitant. Although only about one-third of the forest area can be called regular timber land, Canada possesses an enormous forest wealth, with which she might supply permanently nearly all other countries deficient in material, if the governing bodies in the several provinces would only determine to stop the present fearful waste caused by axe and fire and to introduce a regular system of management. As matters stand, the supplies of the most valuable timber of Canada, the white or Weymouth pine, are nearly exhausted, the great stores of spruce in the eastern provinces are being rapidly destroyed and the forests of Douglas fir in the western provinces have been attacked for export to the United States and to other countries.

Taking the remaining stocks of the whole earth together, it may be said that a sufficient quantity of hardwoods is available, but the only countries which are able to supply coniferous timber for export on a considerable scale are

Russia, Sweden, Norway, Austria and Canada. As these countries have practically to supply the rest of the world and as the management of their forests is far from satisfactory, the question of supplying light pine and fir timber, which forms the very staff of life of the wood industries, must become a very serious matter before many years have passed.

Unmistakable signs of the coming crisis are everywhere visible to all who wish to see and it is difficult to over-state the gravity of the problem, when it is remembered, for instance, that 87% of all the timber imported into Great Britain consists of light pine and fir and that most of the other importing countries are similarly situated.

In some of these countries little or no room exists for the extension of woodland, but this statement does not apply to Great Britain and Ireland, which contain upwards of 12,000,000 acres of waste land and 12, 500,000 acres of mountain and heath land used for light grazing. Onefourth of that area, if put under forest, would produce all the timber now imported which can be grown in Britain, that is to say, about 95% of the total.

Forest Management

In early times there was practically no forest management. As long as the forests occupied considerable areas, their produce was looked upon as the free gift of nature, like air and water; men took it, used it and even destroyed it without let or hindrance. With the gradual increase of population and the consequent reduction of the forest area, proprietary ideas developed; people claimed the ownership of certain forests and proceeded to protect them against outsiders.

Subsequently the law of the country was called in to help in protection, leading to the promulgation of special forest laws. By degrees it was found that mere protection was not sufficient and that steps must be taken to enforce a more judicious treatment, as well as to limit the removal of timber to what the forests were capable of producing permanently.

The teaching of natural science and of political economy was brought to bear upon the subject, so that now forestry has become a special science. This is recognized in many countries, amongst which Germany stands first, closely followed by France, Austria, Denmark and Belgium. Of non-European countries the palm belongs to British India and then follow Ceylon, the Malay States, the Cape of Good Hope and Japan.

The United States of America have also turned their attention to the subject. Most of the British colonies are, in this respect, as yet in a backward state and the matter has still to be fought out in Great Britain and Ireland, though many writers have urged the importance of the question upon the public and the government. There can be no doubt that all civilized countries must, sooner or later, adopt a rational and systematic treatment of their forests.

For details as to the separate countries, see the articles under the country headings; in this article only some of the more important countries are dealt with, in so far as the history of their forestry is important. A few notes on Germany and France will be given, because in these countries forest management has been brought to highest perfection; Italy is mentioned, because she has allowed her forests to be destroyed; and a short description of forestry in the United Kingdom and in India follows.

The winters being long and severe, an abundant supply of fuel is almost as essential as a sufficient supply of food. This necessity has led, along with a passion for the chase, to the preservation of forests and to the establishment of an admirable system of forest cultivation, almost as carefully conducted as field tillage.

The Black Forest stretches the whole length of the grand-duchy of Baden and part of the kingdom of Wurttemberg, from the Neckar to Basel and the Lake of Constance. The vegetation resembles that of the Vosges; forests of spruce, silver fir, Scotch pine and, mingled with birches, beech and oak, are the chief woods met with. Until comparatively recent times large quantities of timber derived from these forests were floated down the Rhine to Holland and also shipped to England.

Now the greater part of it is used locally for construction, or it is converted into paper pulp. In the grand-duchy of Hesse the Odenwald range of mountains, stretching between the Main and the Neckar, contains the chief supply of timber. In the province of Nassau there are the large wooded tracts of the Taunus mountain range and the Westerwald.

In Rhenish Prussia valuable forests lie partly in the Eifel, on the borders of Belgium and on the mountains overhanging the Upper Moselle, but they do not furnish such stately trees as the Black Forest and the Odenwald. The Spessart, near Aschaffenburg in Bavaria, is one of the most extensive forests of middle Germany, containing large masses of fine oak and beech, with plantations of coniferous trees, such as spruce, Scotch pine and silver fir.

Bavaria possesses other fine forest tracts, such as the Baierischewald on the Bohemian frontier, the Kranzberg near Munich and the Frankenwald in the north of the kingdom. North Germany has extensive forests on the Harz and Thuringian Mountains, while in East Prussia large tracts of flat ground are covered with Scotch pine, spruce, oak and beech.

Every German state has its forest organization: In Prussia the department is presided over by the Oberland Forstmeister at Berlin, while each province, or part of a province, has an Oberforstmeister, under whom a number of OberfOrsters administrate the state and communal forests. These, again, are assisted by a lower class of officials called Forsters. The Oberf6rsters throughout Germany are educated at special schools of forestry, of which in 1909 the following nine existed: In Prussia: at Eberswalde and Munden.

The National Forest Policy

The forest policy of the United States may be said to have had its origin in 1799 in the enactment of a law which authorized the purchase of timber suitable for the use of the navy, or of land upon which such timber was growing. It is true that laws were in force under the early governments of Massachusetts, New Jersey and other colonies, providing for the care and protection of forest interests in various ways, but these laws were distinctly survivals of tendencies acquired in Europe and for the most part of little use.

It was not until the apparent approach of a dangerous shortage in certain timber supplies that the first real step in forest policy was taken by the United States. Successive laws passed from 1817 to 1831 strove to give larger effect to the original enactment, but without permanent influence towards the preservation of the live oak, which was the object in view. A long period of inaction followed these early measures. *The unshaded areas are treeless, except along the Streams* 1831 the solicitor of the treasury assumed a partial responsibility for the care and protection of the public timber lands and in 18 55 this duty was transferred to the commissioner of the general land office in the Department of the Interior.

The effect of these changes upon forest protection was unimportant. When, however, at the close of the Civil War railway building in the United States took on an unparalleled activity, the destruction of forests by fire and the axe increased in a corresponding ratio and public sentiment began to take alarm. Action by several of the states slightly preceded that of the Federal government, but in 1876 Congress, acting under the inspiration of a memorial from the American Association for the Advancement of Science, authorized the appointment of an officer under the commissioner of agriculture, to collect and distribute information upon forest matters.

As the railways advanced into the treeless interior, public interest in tree-planting became keen. In 1873 Congress passed and later amended and repealed the timber culture acts, which granted homesteads on the treeless public lands to settlers who planted one-fourth of their entries with trees. Though these measures were not successful in themselves they directed attention towards forestry. The act which repealed them in 1891 contained a clause which lies at the foundation of the present forest policy of the United States.

By it the president was authorized to set aside " any part of the public lands wholly or in part covered with timber or undergrowth, whether of commercial value or not, as public reservations and the President shall, by public proclamation, declare the establishment of such reservations and the limits thereof." Some eighteen million :acres had been proclaimed as reservations at the time when, in 1896, the National Academy of Sciences was asked by the secretary of the interior to make an investigation and report upon " the inauguration of a rational forest policy for the forest lands of the United States."

Upon the recommendation of a commission named by the Academy, President Cleveland established more than twenty-one million acres of new reserves on the 22nd of February 1897. His action was widely misunderstood and attacked, but it awakened a public interest in forest questions without which the rapid progress of forestry in the United States since that time could never have been made.

Within a few months after the proclamation of the Cleveland reserves the present national forest policy took definite shape. Under this policy the national government holds and manages, in the common interest of all users of the forests or its products, such portions of the public lands as have been set aside by presidential proclamation in accordance with the act of 1891.

These lands are held against private acquisition under the Homestead Act (except as to agricultural lands as hereafter mentioned), the Timber and Stone Act and other laws under which the United States disposes of its unappropriated public domain, but not against private acquisition under the Mineral Land Laws. They are selected from lands believed to be more valuable for forest purposes than for agriculture and are managed with the purpose of securing from them the best and largest possible returns, present and future, whether in the form of water for irrigation or power, of timber, of forage for stock, or of any other beneficial product. The aggregate area of the reserves, or national forests, has been steadily increased until they now include nearly all the timber lands left of the public domain.

The general lines of this policy were in part laid down by the commission already mentioned, in its report submitted to the secretary of the interior, May 1, 1897 and by the act of June 4, 1897, which was largely shaped by the work of the commission. Until this act was passed the national forests had been in theory closed against any form of use; nor had the possibility of securing forest preservation by wise use received much thought from those who had favoured their creation. Such a state of affairs could not continue.

Before long public opinion would have forced the opening to use of the resources thus :arbitrarily locked up and in the absence of any administrative system providing for conservative use, the national forests would inevitably have been abolished and the whole policy of government forest holdings would have ceased.

The act of June 4, 1897 was therefore of the first importance. This act conferred upon the secretary of the interior general powers for the proper management of the national forests through the general land office of his department. It provided for the designation and sale of dead, mature and large timber; authorized the secretary to permit free use of timber in small quantities by settlers, miners and residents; empowered him to "make such rules and regulations and establish such service as will insure the objects of such reservations, namely, to regulate their occupancy and use and to preserve the forests thereon from destruction "; and made violation of the act or of such rules and regulations a misdemeanour. The statute limited the power to establish forest reservations to the purpose of improving and protecting the forest, securing favourable conditions of water flows and furnishing a continuous supply of timber for the use and necessities of citizens of the United States.

Lands found, upon due examination, to be more valuable for other purposes than for forest uses might be eliminated from any reservation and all mineral lands within the reservations were left open to private appropriation under the mineral laws. The rights of settlers and claimants were safeguarded and civil and criminal jurisdiction, except so far as the punishment of offences against the United States in the reservations was concerned, was reserved to the States.

While the administration of the national forests was entrusted to the general land office, the same act assigned the surveying and mapping of them to the United States Geological Survey, which has published descriptions and maps of some of the more important.

No attempt was made in the general land office to develop a technical forest service. There were, indeed, at the time of passage of the act, less than ten trained foresters in the United States, no means of training more and very little conception of what forestry actually meant. The purpose of the administration was therefore mainly protection against trespass and fire, particularly the latter.

Regulations were made giving effect to the provisions of the act of June 4, set forth above, but in the absence of technical knowledge as to what might safely be done, the tendency was rather to restrict than to extend the use of the forest. Meanwhile, however, there was rapidly developing in another branch of the government service an organization qualified for actual forest management.

One year after the passage of the act of June 4, 1897, the division of forestry in the Department of Agriculture ceased to be merely a bureau of information and became an active agency for introducing the actual practice of forestry among private owners and for conducting the investigations upon which a sound American forest practice could be based. The work awakened great interest among forest owners and exerted a powerful educational influence upon the country at large.

The division extended its work and became the Bureau of Forestry. It drew into its employment for a time nearly all the men who were preparing themselves in increasing numbers (at first abroad, then in the newly-founded schools in the United States) for the profession of forestry and was soon recognized as qualified to speak authoritatively on technical questions connected with the administration of the national forests.

This led to a request from the secretary of the interior for the advice of the bureau on such questions. Working plans were accordingly undertaken for a number of the forests. The general land office, however, was not ready to attempt active forest management. Though some timber was sold and the grazing of stock regulated to some extent, the main object of the land office administration continued to be protection against fire. Many of the regulations which it made could not be enforced.

The disadvantages of dispersal of the Federal government forest work among three separate agencies grew more and more apparent, until, on the 1st of February 1905, control of the 63,000,000 acres of forest reserves which up to that time had been set aside was transferred from the general land office to the Bureau of Forestry. In recognition of its new duties the designation of the bureau became the Forest Service.

Other provisions of the act which affected the transfer were that forest supervisors and rangers should be selected, so far as possible, from qualified citizens of the state or territory in which each forest was situated and that all money received from the sale of any products or the use of any land or resources of the national forests should be covered into the treasury and constitute a special fund for their protection, administration, improvement and extension.

Five days later a statute gave forest officers the power to arrest trespassers; and on the 3rd of March the lieu land selection law was repealed. This law had opened the way for grave abuses through the exchange of worthless land by private owners within the forests for an equal area of valuable timber lands outside.

The law has been modified since by the change of the old name " Forest Reserves " to " National Forests." The act of June iz, 1906, opened to homestead entry lands within national forests found by examination to be chiefly valuable for agriculture. The administration and improvement of the national forests are now provided for directly by congressional appropriation. The power to create national forests conferred on the president by the act of March 1891 has been repealed for the states of Washington, Oregon, Idaho, Montana, Wyoming and Colorado, but for no others.

The Forest Service began in earnest the development of all the resources of the national forests. Mature timber was sold wherever there was a demand for it and the permanent welfare of the forests and protection of the streams permitted, but always so as to prevent waste, guard against fire, protect young growth and ensure reproduction. Regulations were adopted which allowed small sales to be made without formality or delay, secured for the government the full value of timber sold and eliminated unnecessary routine.

Care was taken to safeguard the interests of the government and provide for the maintenance of good technical standards. The conduct of local business was entrusted to local officers. Large transactions with general policies were controlled from Washington, but with careful 2 = *District Boundaries and Numbers* provision for first-hand knowledge and close touch with the work in the field.

Business efficiency and the convenience of the public were carefully studied. In short, an organization was created capable of handling safely, speedily and satisfactorily the complex business of making useful a forest property of vast extent, scattered through sixteen different states of an aggregate area of over 1,50o,000 sq. m. and with a population of 9,000,000.

The growth since the 1st of July 1897 of the area of the national forests, of the expenditures of the government for forestry and of the receipts from the national forests, is shown by the statement which follows. Though the act of June 4, 1897, became effective immediately upon its passage, the fiscal year 1899 was the first of actual administration, because the first for which Congress made the appropriation necessary to carry out the law.

Forest is ripe for the axe, the demand is strong and control by trained men makes it safe to cut more freely. The increase is marked both in small and in large sales, but a score of sales for less than $5000 are made against one for more. The total cut is still far below the annual increment of the forests. As the demand grows restrictions must increase in order to husband the present supply until the next crop matures.

- The stumpage price would seem on the face of the figures to have risen from about one dollar to more than three dollars per thousand board-feet. The receipts, however, for any one year are not exclusively for the timber cut in that year, since payments are made in advance. In the year 1907 the average price obtained was something less than $2.50 per thousand. It is therefore true that stumpage prices have risen greatly, although conditions new to the American lumbermen are im *Area of National Forests, Annual Expenditures of the Federal Government for Forestry and National Forest Administration and Receipts from National Forests, 1898-1909* Until 1906, the sole source of receipts was the sale of timber. In the fiscal year 1907, however, timber sales furnished less than half the receipts. The following statement concerning the timber sales of the fiscal years 1904-1907 will serve to bring out the change that followed the transfer of control to the forest service in the midst of the fiscal year 1905:
  - A large excess each year in the amount of timber sold over that cut and paid for;
  - Nine times as much timber sold at the end of the four-year period as at the beginning and three times as much cut;
  - A much higher price obtained per thousand board-feet at the end of the period than at the beginning. Each of these matters calls for comment. The sales are of stumpage only; the government does no logging on its own account.
- More timber is sold each year than is cut and paid for, because many of the sales extend over several years. With increasing sales the

amount sold each year for future removal has exceeded the amount to be removed during that year under sales of earlier years. Large sales covering a term of years are made because the national forests contain much overmature timber, which needs removal, but which is frequently too inaccessible to be saleable in small amounts. To prevent speculation the time allowed for cutting is never more than five years and cutting must begin at once and be continued steadily.

- The volume of sales has increased rapidly because much 1 The United States fiscal year ends June 30 and receives its designation from the calendar year in which it terminates. Thus, the fiscal year 1898 is the year July 1, 1897 - June 30, 1898.

Administration transferred to Bureau of Forestry, February I, 1905. Full utilization of all merchantable material, care of young growth in felling and logging and the piling of brush, to be subsequently burned by the forest officers if burning is necessary, are among these conditions.

Timber to be cut must first be marked by the forest officers. Sales of more than $100 in value are made only after public advertisement.

Only the simplest forms of silviculture have as yet been introduced. The vast area of the national forests, the comparatively sparse population of the West, the rough and broken character of the forests themselves and the newness of the problems which their management presents, make the general application of intensive methods for the present impracticable.

Natural reproduction is secured. The selection system is most used, often under the rough and ready method of an approximate diameter limit, with the reservation of seed trees where needed.

The tendency, however, is strongly towards a more flexible and effective application of the selection principle, as a better trained field force is developed and as market conditions improve.

One conspicuous achievement was the reduction of loss by fires on the national forests. During the unusually dry season of 1905 there were only eight fires of any importance and the area burned over amounted only to about 16 of 1% of the total area. In 1906 about 12 of 1% was burned. This was accomplished by efficient patrol, co-operation of the public and by preventive measures, such as piling and burning the brush on cut-over areas.

Since the beginning of 1906 the largest source of income from the national forests was their use for grazing. Stock-raising is one of the most important industries of the West. Formerly cattle and sheep grazed freely on all parts of the public domain. In the early days of the national forests the wisdom of permitting any grazing at all upon them was sharply questioned.

Unrestricted grazing had led to friction between individuals, the deterioration of much of the range through overstocking and serious injury to the forests and stream flow. The forests of the West, however, are largely of open growth and contain many grassy parks, the results of old fires and many high mountain meadows.

Under proper regulations the grass and other forage plants which they produce in great quantity can be used without detriment to the forests themselves and with great benefit to the stock industry, which often can find summer pasturage nowhere else.

Except in southern California grazing is now permitted on all national forests unless the watersheds furnish water for domestic use; but the time of entering and leaving, the number of head to be grazed by each applicant and the part of the range to be occupied are carefully prescribed. Planted areas and cut-over areas are closed to stock until the young growth is safe from harm and goats are allowed only in the brushland of the foothills.

The results of regulation, in addition to the protection of forest growth and streams, are the prevention of disputes, improved range, better stock, stable conditions in the stock industry and the best use of the range in the interest of progress and development. The first right to graze stock on the forests is given to residents, small owners and those who have used the range before. Thus the crowding out of the weaker by the stronger and of the settler by the roving outsider has been stopped. In 1906 the forest service began to impose a moderate charge for the use of the national forest range.

The following statement shows the amount of stock grazed on the national forests 1904-09 and the receipts for the grazing charge: - A work of enormous magnitude which has now begun is planting on the national forests. At present, with low stumpage prices and incomplete utilization of forest products, clear cutting with subsequent planting is not practicable.

There are, however, many million acres of denuded land within the national forests which require planting. Such planting is still confined chiefly to watersheds which supply cities and towns with water. The first planting was done in 1892, in California. Since then similar work has been done on city watersheds in Colorado, Utah, Idaho and New Mexico. Other plantations are in the Black Hills national forest, where large areas of cut-over and burned-over land are entirely without seed trees and in the sandhill region of Nebraska. Up to 1908 about 2,000,000 seedlings had been planted, on over 2000 acres - a small beginning, but the work was entirely new and presented many hard problems.

The nursery operations of the forest service are concentrated at seven stations, located in southern California, Nebraska, Colorado, New Mexico Utah and Idaho, where stock is raised for local planting and for shipment elsewhere. These nurseries are small. Their annual productive capacity is between 8,000,000 and seedlings. Each nursery is practically an experimental forest-planting station, at which a large variety of species are grown and various methods are tried.

The organization of the administrative work of the national forests is by single forests. On the 1st of January 1908 the total number of forests was 165 with a total area of 162,023,190 acres. In charge of each forest is a forest supervisor. Under the supervisors are forest rangers and forest guards, whose duties include patrol, marking timber and scaling logs, enforcing the regulations and conducting some of the minor business arising from the use of the forests. Guards are temporary employes; rangers are employed by the year. The supervisors report directly to and receive instructions from the central office at Washington.

In this office there are four branches - operation, grazing, silviculture and products - each of which directs that part of the work which belongs to it, dealing directly with the supervisor.

For inspection purposes, however, the forests are separated into six districts, in each of which is located a chief inspector with a corps of assistants. The inspectors are without administrative authority, but assist by their counsel the supervisors and through inspection reports keep the Washington office informed of the condition of all lines of administrative work in progress. Administrative officers alternate frequently between field and office duties.

The number of forest officers in the several grades on the 1st of January 1908 were: 6 chief inspectors, 26 inspectors, 106 forest supervisors, 41 deputy forest supervisors, 820 forest rangers and 283 forest guards. The total number of employes of the forest service on the same date, including the clerical force, was 2034.

Besides the administration of the national forests, the forest service conducts general investigations, carries on an extensive educational work and co-operates with private owners who contemplate forest management upon their own tracts. This last work is undertaken because of the need of bringing forestry into practice, the lack of trained foresters outside of the employ of the government and the lack of information as to how to apply forestry and what returns may be obtained. Co-operation takes the form of advice upon the ground and, on occasion, of the making of working plans.

The educational work of the service is performed chiefly through publications, the purpose of which is to spread very widely a knowledge of the importance of forestry to the nation and of the principles upon which its practice rests.

The investigations which the service conducts extend from studies. of the natural distribution and classification of American forests. and of their varied silvicultural problems to statistics of lumber production and laboratory researches which bear upon the economical utilization of forest products. As examples of these researches may be mentioned tests of the strength of timber, studies of the preservative treatment of wood for various uses, wood-pulp investigations and studies in wood chemistry.

Forestry Among the States

Among the states forestry has hardly reached the stage of practical application on the ground. New York holds 1,500,00o acres of forest land. It has a commission to care for its forest preserve and to protect the forest land throughout the state from fire. The constitution of the state, however, prohibits the cutting of timber on state land and thus confines the work entirely to protection of the forest and to the planting of waste areas. Pennsylvania is at present showing the most efficient activity in working out a forest policy. It has state forests of 820,000 acres, a good fire law more and more satisfactorily enforced and eight nurseries for growing planting material. In 1905, 160,000 white pine seedlings were set out. It has also a school for forest rangers, to be employed on the state forests and it has just established a state professional school of forestry.

Twenty-six of the states have regularly appointed forest officers, six have carried on studies of forest conditions in co-operation with the forest service and there is scarcely one which is not actively interested in forestry. Laws, generally good, to prevent damage from forest fires, have been enacted by practically all the states, but their enforcement has unfortunately been lax.

Public sentiment, however, is making rapid progress. Among the best laws are those of Maine, New Hampshire, Minnesota, New York, Pennsylvania and Wisconsin. The New York law, for example, provides for the appointment of one or more firewardens in each town of the counties in which damage by fire is especially to be feared. In other counties supervisors of towns are *ex-officio* fire-wardens. A chief fire-warden has general supervision of their work.

The wardens, half of the cost of whose services is paid by the state, receive compensation only for the time actually employed in fighting fires. They may command the service of any citizen to assist them. Setting fire to woods or waste lands belonging to the state or to another, if such fire results in loss, is punishable by a fine not exceeding $250 or imprisonment not exceeding one year, or both and damages are provided for the person injured. Since fire is beyond question the most dangerous enemy of forests in the United States, the measures taken against it are of vital importance.

The amount of forest land held by the different states and by the territory of Hawaii: A *rea of State Forest Reservations, 1907. Forestry on Private Lands. The* practice of forestry among private owners is of old date. One of the earliest instances was that of Jared Eliot, who, in 1730, began the systematic cutting of timber land to supply charcoal for an iron furnace at Old Salisbury, Connecticut. The successful planting of waste lands with timber trees in Massachusetts dates from about ten years later. But such examples were comparatively rare until recent times.

At present the intelligent harvesting of timber with a view to successive crops, which is forestry, is much more common than is usually supposed. Among farmers it is especially frequent. It was begun among lumbermen by the late E. S. Coe, of Bangor, Maine, who made a practice of restricting the cut of spruce from his forests to trees ro, 12 or sometimes even 14 in. in diameter, with the result that much of his land yielded, during his life, a second crop as plentiful as the first.

Many owners of spruce lands have followed his example, but until very recently without improving upon it. Systematic forestry on a large scale among lumbermen was begun in the Adirondacks during the summer of 1898 on the lands of Dr W. S. Webb and Hon. W. C. Whitney, of a combined area of over ioo,000 acres, under the superintendence of the then Division of Forestry. In these forests spruce, maple, beech and birch predominate, but the spruce alone is at present of tht first commercial importance. The treatment is a form of the selection system. Under it a second crop of equal yield would be ripe for the axe in thirtyfive years. Spruce and pine are the only trees cut. The work had been executed, at least up to the year 1902, with great satisfaction to the owners and the lumbering contractors, as well as to the decided benefit of the forest. The lumbering is regulated by the following rules and competent inspectors are employed to see that they are rightly carried out:

- No trees shall be cut which are not marked.
- All trees marked shall be cut.
- No trees shall be left lodged in the woods and none shall be overlooked by the skidders or haulers.
- All merchantable logs which are as large as 6 in. in diameter at the small end must be utilized.
- No stumps shall be cut more than 6 in. higher than the stump is wide.
- No spruce shall be used for bridges, corduroy, skids, slides, or for any purpose except building camps, dams or booms, unless it is absolutely necessary on account of lack of other timber.
- All merchantable spruce used for skidways must be cut into logs and hauled out.
- Contractors must not do any unnecessary damage to young growth in lumbering; and if any is done, they must discharge the men who did it.

These two instances of forestry have been most useful and effective among lumbermen and other owners of forest land in the north-east. Among those which have followed their example are the Berlin Mills Paper Company in northern New Hampshire,, the Cleveland Cliffs Iron Company in northern Michigan and the Delaware and Hudson Railroad Company in New York, all of which have employed professional foresters.

The most notable instance of forestry in the south is on the estate of George W. Vanderbilt at Biltmore, N.C. This was the first case of systematic forestry under regular working plans in the United States. It was begun in 1891 on about 4000 acres and has since been extended until it now covers about ioo,000 acres. A professional forester with a corps of trained rangers under him is in charge of the work. The Pennsylvania Railroad has recently employed a trained forester and several assistants and has undertaken systematic forestry on a large scale.

The effect of the work of the forest service in assisting private owners is evidenced by the fact that down to the year 1908 670 wood lots and timber tracts had been examined by agents of the forest service, of which 250 were tracts over 400 acres in extent and planting plans had been made for 436 owners covering a total area of 80,000 acres. Expert advice is also given to wood lot owners upon application by many of the state foresters.

American Practice

The conditions under which forestry is practised in Europe and in America differ so widely that rules which are received as axiomatic in the one must often be rejected in the other. Among these conditions in America are the highly developed and specialized methods and machinery of lumbering, the greater facilities for transportation and consequent greater mobility of the lumber trade, the vast number of small holdings of forest land and the enormous supply of low-grade wood in the timbered regions.

High taxes on forest properties, cut-over as well as virgin, notably in the north-western pineries and the firmly established habits of lumbermen, are factors of great importance. From these and other considerations it follows that such generally accepted essentials of European methods of forestry as a sustained annual yield, a permanent force of forest labourers, a permanent road system and the like, are in most cases utterly inapplicable in the United States at the present day in private forestry. Methods of forest management, to find acceptance, must there conform as closely as possible to existing methods of lumbering. Rules of marked simplicity, the observance of which will yet secure the safety of the forest, must open the way for more refined methods in the future.

For the present a periodic or irregular yield, temporary means of transport, constantly changing crews and an almost total ignorance of the silvics of all but a few of the most important trees - all combine to enforce the simplest silvicultural treatment and the utmost concentration of purpose on the two main objects of forestry, which are the production of a net revenue and the perpetuation of the forest. Such concentration has been followed in practice by complete success.

The forests with which the American forester deals are rich in species,usually endowed with abundant powers of reproduction and, over a large part of their range, greatly dependent for their composition and general character upon the action of forest fires. Of the commercially valuable trees there may be said to be, in round numbers, a hundred out of a total forest flora of about 500 species, but many trees not yet of importance in the lumber trade will become so hereafter, as has already happened in many cases.

The attention of the forester must usually be concentrated upon the growth and reproduction of a single species and never of more than a very few. Thus the silvicultural problems which must be solved in the practice of forestry in America are fortunately less complicated than the presence of so many kinds of trees in forests of such diverse types would naturally seem to indicate.

The forest fire problem is one of the most difficult with which the American forester has to deal. It is probable that forest fires have had more to do with the character and distribution of forests in America than any other factor except rainfall. With an annual range over thousands of square miles, in many portions of the United States they occur regularly year after year on the same ground. Trees whose thick bark or abundant seeding gives them peculiar powers of resistance, frequently owe their exclusive possessions of vast areas purely to the action of fire.

On the economic side fire is equally influential. The probability, or often the practical certainty, of fire after the first cut, commonly determines lumbermen to leave no merchantable tree standing. Forest fires are thus the most effective barriers to the introduction of forestry.

Excessive taxation of timber land is another of almost equal effect. Because of it lumbermen hasten to cut and afterwards often to abandon, lands which they cannot afford to hold. This evil, which only the progress of public sentiment can control, is especially prevalent in certain portions of the white pine belt.

Forest Associations

Public sentiment in favour of the protection of forests is now widespread and increasingly effective throughout the United States. As the general understanding of the objects and methods of forestry becomes clearer, the tendency, formerly very marked, to confound ornamental tree planting and botanical matters with forestry proper is rapidly growing less. At the same time, the number and activity of associations dealing with forest matters is increasing with notable rapidity. There are now about thirty such associations in the United States. One of these, the Society of American Foresters, is composed exclusively of professional foresters. The American Forestry Association is the oldest and largest.

It has been influential in preparing the ground work of popular interest in forestry and especially in advocating and securing the adoption of the federal forest reservation policy, the most important step yet taken by the national government. It publishes as its organ a monthly magazine called *Forestry and Irrigation.* The Pennsylvania Forestry Association has been instrumental in placing that state in the forefront of forest progress. Its organ is a bi-monthly publication called *Forest Leaves.*

Other states which have associations or societies of special influence in forest matters are California, Massachusetts, Minnesota, Colorado, New Hampshire, Georgia and Oregon. Arbor Day, instituted in Nebraska in 1872 as a day for shade-tree planting by farmers who had settled on the treeless prairies, has been taken up as a means of interesting school children in the planting of trees and has spread until it is now observed in every state and territory. It continues to serve an admirable purpose.

Lumbering

All the operations of the lumber trade in the United States are controlled and to no small degree determined, by the peculiar unit of measure which has been adopted. This unit, the boardfoot, is generally defined as a board one foot long, one foot wide and one inch thick, but in reality it is equivalent to 14 4 cub. in. of manufactured lumber in any form. To purchase logs by this measure one must first know about what each log will yield in one-inch boards.

For this purpose a scale or table is used, which gives the contents of logs of various diameters and lengths in board feet. Under such a standard the purchaser pays for nothing but the saleable lumber in each log, the inevitable waste in slabs and sawdust costing him nothing.

In addition to this amount, an immense quantity of wood is used each year for fuel, posts and other domestic purposes and the total annual consumption is not less than 20 billion cub. ft. The years 1890 to 1906 were marked by rapid changes in the rank of the important timber trees with reference to the amount of timber cut and a shifting of the important centres of production.

Among coniferous trees, white pine has yielded successively to yellow pine and Douglas fir, while the scene of greatest activity has shifted from the Northern forest to the. Southern and from there is rapidly shifting to the Pacific Coast. The total cut of coniferous lumber has increased steadily, but that of the hardwoods is falling off and in 1906 it was 15% less than in 1899, while inferior hardwoods are gradually assuming more and more importance and the scene of greatest activity has passed from the middle west to the south and the Appalachian region. *Conifers. The* coniferous supply of the country is derived from four forest regions:

- The Northern forest;
- The Southern forest;
- The Pacific Coast forest;
- The Rocky Mountain forest.

The Northern forest was long the chief source of the coniferous lumber production in the United States. The principal timber tree of this region is the white pine, usually known in Europe as the Weymouth pine. It has an average height when mature of i to ft., with a diameter a little less than 3 ft., but the virgin timber is approaching exhaustion. White pine was one of the first trees to be cut extensively in the United States and Maine, the pine tree state, was at first the centre of production. In 1851 the cut of white pine on the Penobscot river was 144 million ft., that of spruce 14 million and of hemlock 11 million.

Thirty years later the pine cut had sunk to 23 million, spruce had risen to 118 million and hemlock had passed pine by a million feet. Meanwhile, the centre of production had passed from the north woods to the Lake States and for many years this region was the scene of the most vigorous lumbering activity in the world. Woods waste includes tops, stumps, cull logs and butts, but does not include defective trees left or trees used for road purposes. Mill waste includes bark, kerf, slabs and edgings.

The spruce (*Picea rubens*) is used chiefly for lumber, but it is in large and increasing demand in the manufacture of paper pulp. For the latter purpose hemlock, poplar and several other woods are also employed, but on a smaller scale. The total consumption of wood for paper in the United States for 1906 was 3,660,000 cords, of which 2,500,000 was spruce. Of this, however, 720,000 cords were imported from Canada.

- The chief product of the Southern forest is the yellow pine. This is the collective term for the longleaf, shortleaf, loblolly and Cuban pines. Of these the longleaf pine, called pitch-pine in Europe, is the most important. Its timber is probably superior in strength and durability to that of any other member of the genus *Pinus* and in addition to its value as a timber tree it is the source of naval stores in the United States. The average size of the mature longleaf pine is 90 ft. in height and 20 in. in diameter. Shortleaf and loblolly are other important members of this group. Their wood very closely resembles that of the longleaf pine and is often difficult to distinguish from it. The trees are also of about the same size and height. Loblolly is, however, of more rapid growth. The total cut of yellow pine in 1906 was 11,661,000,000 board ft.; it has perhaps not yet reached its maximum, but is certainly near it.

  Another important coniferous tree of the Southern forest is the bald cypress, which grows in the swamps. The cut in 1906 was 839,000,000 board ft., a gain of 69% over 1899.
- But the great supply of coniferous timber is now on the Pacific Coast. The Douglas fir, also known as Douglas spruce, red fir and Oregon pine, is the foremost tree in Oregon and Washington and the redwood in California. When mature the Douglas fir averages 200 ft. in height and 4 ft. in diameter and the redwood 225 ft. in height and 8 ft. in diameter. Other important trees of the Pacific Coast are sugar pine, western red cedar, western larch, Sitka spruce, western hemlock and western yellow pine.

  Logging on the Pacific Coast is characterized by the use of powerful machinery and by extreme skill in handling enormous weights. This is especially true in California, where the logs of redwood and of the big tree are often more than f ft. in diameter. Logging is usually done by wire cables operated by donkey-engines. The journey to the mill is usually by rail. The mills are often of great size, built on piles over tide water and so arranged that their product is delivered directly from the saws and dry kilns to vessels moored alongside. The products of the Pacific Coast forest make their way over land to the markets of the central and eastern states and into foreign markets. Among the lumber-producing states, Washington has in seven years jumped from fifth place to first and its output has increased from 1,428,000,000 board ft. in 1899 to 4,305,000,000 ft. in 1906. Oregon and California have increased their output from 734,000,000 each in 1899 to 1,605,000,000 and 1,349,000,000 ft. respectively in 1906. Of the total output of these three states (7,259,000,000 ft.) 4,880,000,000 ft. is Douglas fir and 660,000,000 redwood.

- The important lumber trees of the Rocky Mountain forest are the western yellow pine, the lodgepole pine, the Douglas fir and the Engelmann spruce. The Douglas fir, here extremely variable in size and value, reaches in this region average dimensions of perhaps 80 ft. in height by 2 ft. in diameter, the western yellow pine 90 ft. by 3 ft. and the Engelmann spruce 60 ft. by 2 ft. Mining, railroad and domestic uses chiefly absorb the annual timber product, which is considerable in quantity and of vast importance to the local population.

RIGHTS OVER FOREST RESOURCES

The tribal people are facing serious problems with regard to utilisation and rights over forest resources. Due to the increasing pressure on forests by various interest groups, there is a corresponding pressure on the tribals to reduce their dependency on forests. This is creating serious situations of conflict, as tribal life is symbiotic with land and forests and their livelihood and culture are based on their relationship with the natural wealth around them. The tribals are being harassed for using forestlands and being evicted in many places. Such reports have come from places like Khammam, Visakhapatnam, Vizianagaram, Adilabad, Srisailam and other places. In Khammam in one particular village, the forest and police departments allegedly branded the tribals on their shoulders as an indication that they were destroying the forests. There are ambiguities in forest-revenue land demarcation. In some places like Nellore district, a lot of land on Velugonda hills is indicated as 'poramboku' in revenue records and as RF as per the forest department. These lands do not have any forest growth. The FD is taking up palm oil plantation in these RF lands (which is not a forest species). However, landless tribals are booked in criminal cases or prohibited from using these lands for agricultural cultivation. These lands should be given to tribals with pattas for cultivation. In some areas like in Visakhapatnam district, the lack of clear forest boundaries is making tribals vulnerable to the exploitation of both the forest and revenue departments. A joint survey and demarcation of boundaries by both departments should be immediately taken up to arrest these conflicts. Such Joint surveys need to involve the villagers at various levels. The Forest-Revenue Boundary dispute is a perpetual problem in Adilabad and Warangal districts, leading to booking of cases by the Forest department and tension in these tribal villages.

In Visakhapatnam Agency and Nallamala areas, there is the unique problem of Enclosure Villages. There were many tribal villages that were not enumerated in the forest surveys. Due to such sheer negligence, the villages were not given revenue status and to this day, they do not have pattas for their lands. They face constant harassment from local forest officials, as they do not possess land records. Recognizing these enclosure villages and issuing pattas to tribals should immediately resolve this problem. In Buttapur (Adilabad district) and in Nellore Dt. (Yanadis), the tribals were given lands decades ago under the social forestry scheme and are cultivating there. But due to lack of pattas, they are being harassed by the police and forest departments and also do not have access to bank loans as they cannot prove their ownership. These tribals have to be given pattas as promised.

Attacks on tribals, their properties and livestock by wildlife are not compensated by the Forest Department. Several cases are pending where tribals have been either killed or disabled and yet have not received any monetary compensation as due to them under the Wildlife Act. The tribals should not be prohibited from entering the forests to collect NTFP for their domestic requirements, like firewood, medicinal herbs, food, or agricultural and housing material.

The tribals should not be defined as ‘encroachers’, as is being projected by the forest department and also as indicated in the Circular of IG Forests, MoEF dated 3rd May 2002. In districts like Visakhapatnam, East Godavari and Vizianagaram, where there is high prevalence of podu cultivation, the tribals are facing threats of eviction from the forest department. There was a notification issued by the government of A.P in 1987 ordering that pre-1980 settlements will not be evicted until further orders. This should be implemented. The JFM programme now renamed as the Community Forest Management programme of the A.P Forest Department has caused grievous violations with regard to tribal rights. One major violation is the displacement of tribals from their podu lands by reclaiming them back into the forests through the JFM programme. The official reports of the forest department and the World Bank (which has funded the project) reveal that 37,000 hectares of forestland has been reclaimed back from the people§. The APFD further states its intention of displacing tribals in G.O. where it calls for voluntary surrender of lands by involving NGOs to motivate people under the CFM programme. The APFD has come up with a proposal for rehabilitation of tribals displaced from their podu lands through a monetary and schematic approach. This should be condemned as it bypasses the issue of eviction and rights of people over their podu lands. Their act of reclaiming lands under JFM programme and the APFD's forestry project under World Bank assistance has to be scrutinized closely by the Commission.

Samata and a few other NGOs raised this issue by writing to the World Bank on its violations of its own Operational Directives 4.2 and 4.3 in its financial support to the APFD. This has led to the Bank's insistence on an R&R policy to be proposed and implemented by the APFD in its CFM programme

In Srisailam area, the Rajiv Gandhi Tiger Sanctuary has led to eviction of tribals from their original homes. They have not been properly settled so far. The concept of EDC (Eco Development Committees) that was introduced is working to the detriment of the tribals, as the objective of this programme was to reduce the forest dependency of the tribals. The tribals are given income-generating programmes and alternate sources of firewood by the forest department. This is an artificial mechanism that is not sustainable in the long term.

In the Srisailam Tiger Sanctuary area, the Chenchus, who are traditional hunter-gatherers, go into the forest everyday for all their needs. They are being harassed by the forest department for trespassing into the sanctuary. The groups who came from the chenchu area felt that the Chenchus should atleast be given identity cards to prevent harassment from forest department and the police department who mistake them for naxalites.

Another serious problem with regard to Sanctuaries is the settlement of people's rights. The Settlement Officer is the Conservator of Forests whereas the forest department is one of the interested parties in the land acquisition for the sanctuary and hence it is not appropriate to make the forest officer the settlement officer. It should be the revenue department (District Collector) who should be authorised to settle the people's rights. G.O.No.112 which calls for involvement of private industries like ITC, Reliance and others for taking up commercial plantations through the VSS should be withdrawn immediately, as this is a backdoor method of allowing private industries to enter forest and tribal lands. Such tripartite agreements between the government, the industries and the tribals can never provide a level playing ground for the tribal people and make them more vulnerable to the exploitation of private industries.

The historical injustices to the tribal people in this region and that of the tribals in southern Orissa due to construction of several 'development' projects and industries like Nalco, HAL, Sileru, Machkund and others which have displaced tribals in large numbers without any rehabilitation, and has forced them to migrate in search of livelihood. They are now being treated as encroachers and criminals in forestlands. This definition should be condemned and Commission should recommend proper mechanisms for resolution of conflicts. The Commission should strongly recommend the orders passed by the Ministry of Environment and Forests in 1990, on the recommendations of the Commissioner of Scheduled Castes and Scheduled Tribes, to ensure that such conflicts are resolved. Infact, the first two orders alone show a way forward to settle disputes concerning forestlands.

All the groups consulted strongly felt that the pressure on forests and the destruction of forests is taking place more due to increase in industries and private commercial activities like Mining, paper-mills, timber smuggling and the growing non-tribal population settling down in Agency areas, than due to the podu cultivation practiced by tribals. Eco-Tourism is also causing destruction of forests. For example, the A.P Tourism Department built huge infrastructure close to Farhanbad, near Mannanur (Srisailam) for Tourism purposes in the middle of the protected area and in Araku, Borra and Anantagiri (Visakhapatnam). While tribals are prohibited from entering these forests for their survival needs, it is totally unjust to allow tourism, which is causing a lot of degradation. Hence, the constitutional provisions of the Fifth Schedule, the Samatha Judgement and the PESA Act should be strictly implemented following the right spirit with regard to industries and non-tribal settlers, which includes the Tourism Department.

The Tourism Department also conducts tours through tribal villages, especially in Chenchu settlements, and exhibits them like museum pieces. The groups felt that this degrading exercise further underscored the lack of empathy among the various government departments towards the tribals.

Because of a shift in the policy of the government, profits to G.C.C. are more important than benefits to tribals. Although there are 32 items in the procurement list of GCC Ltd, it procures a limited number of commodities and at very low prices compared to the market rates. The tribals face several constraints in collection of NTFP and marketing these items because of the monopoly rights enjoyed by GCC Ltd. They do not have the right to sell their produce to private traders even when their rates are higher. Hence, a long-standing demand has been for removal of the monopoly restriction of GCC Ltd and instead a minimum support price should be provided by GCC to protect the tribals from exploitation of traders. The monopoly of forest department over Beedi leaf collection and sale should be removed. The tribals or VSS should be given 100% rights over income from beedi leaf and also the right of sale, with the forest department providing the support price.

NATIONAL FOREST POLICY OF JOINT FOREST MANAGEMENT

One outcome of the National Forest Policy was the launching of Joint Forest Management (JFM) initiatives by all the States and Union Territories. However, there have been no accompanying changes in the legal framework for JFM, and in almost all states it has been implemented through administrative orders, which do not take the existing infrastructure into consideration. Forests continue to be looked upon as the property of the Government and the delegation of management responsibilities to local communities is merely an administrative arrangement, falling short of any real devolution of power to local communities.

As an illustration. Prakash Kashwan has looked into JFM from various points of view and highlights some other problems. One of the most important findings is that JFM has been allowed only in degraded forest, but in order to bring more communities under the ambit of such programme, communities have had to actively pursue degradation of natural forest.

Furthermore, JFM requires the demarcation of boundaries between villages, often for the first time, but poor implementation has resulted in many boundary conflicts.

Joint Forest Management (JFM) in the state of Jharkand;

- JFM in the state of Jharkand is based on the premise that people are to blame for the destruction of the forest. The objective, therefore, is gradually to diminish people's dependence on forests.
- The Resolution to introduce JFM has been made unilaterally by the Forest Department. There is no scope for consultation with the people, no effort at consensus building, and no mechanism for dissemination of information.
- The resources required for villagers' needs are to be extracted from the protected forest according to the micro/annual-working plan. There is no mention of the existing rights of the villagers to use the produce from these forests. This is a gross violation of customary rights.
- A 33% share in net profits was promised. Now the promise is to share 90% of net profits with the forest dwellers. The profits however, are from commercial productions, which are not based on local needs, thus indicating the focus on commercialization of forest. The JFM Resolution in Jharkhand is simply an extension of the undivided Bihar Forest Policy (Jharkhand used to be part of Bihar before it became a separate state), which adopted Joint Forest Management in 1990. Jharkhand Forest Department adopted the JFM Resolution on 27th September 2001
- The Forest Department claims to have formed 2034 Village Forest Management and Conservation Committees and Eco-Development Committees managing an area of 5,589 sq. km. in a remarkably short time span. However, local people emphatically rejected JFM. Apparently, the figure is manipulated and incorrect.

Eviction of communities as Forest Encroachers

Indigenous peoples feel that measures that were put in place in the name of protecting the forest have only deprived them of their resources. The JFM initiatives have failed from the point of view of indigenous peoples, and there has also been widespread eviction of communities from forests. The forest department uses any means available to alienate the forest communities from the forest, reduce their dependence on the forest in the name of forest protection and conservation, and invite private partners and multinational companies to negotiate directly with village committees to access forests for carbon trading, Clean Development Mechanisms and trade in eco-services. The direction of forest management is thus very far from anything that takes traditional forest related knowledge into account or that recognises the rights of indigenous peoples to maintain their traditional forest practices; rather, it aims to separate them from the forests, and sets up an externally imposed process of commodification and marketing of forest goods.

Indigenous Peoples' Rights and Forest Management

The Government of India is not a signatory to the ILO Convention 169 and as such does not recognize indigenous peoples. It does recognise "Scheduled Tribes", but not all indigenous groups are scheduled. Some of the indigenous groups are recognized by the Government of India as Scheduled Tribes and enjoy a degree of special status. Some of the areas where the Scheduled Tribes live have been demarcated as Scheduled areas and under the Indian Constitution Sixth Scheduled provisions, Autonomous District Councils have been set up for local administration of the Scheduled Tribes.

In theory there are some measures to protect the interests of Scheduled Tribes. The Constitution empowers the President of India and state governors to withhold any law considered detrimental to tribal/indigenous peoples' interests in Scheduled Areas. Items V and VI of the Indian Constitution give special privileges to the Scheduled Tribes and the Panchayats Act (Extension to the Scheduled Areas), 1996 (PESA) is also designed to offer some protection. However, in practice most of the laws restrict the rights of and control of forest communities. Specifically, the Indian Forest Act (IFA), Forest Conservation Act (FCA) and Wildlife Protection Act (WLPA) continue to be used to hound forest dwellers.

JFM and CFM are present in most Indian states, but are simply plantation programmes related to timber and other economic products, which have nothing in essence to do with forests. They should include Village Forest Committees directly representing the community living in or using the forest. These programmes are operational to varying degrees in different places with varying degrees of participation of indigenous peoples depending upon the local situation. The involvement of indigenous peoples is focused on providing them with a share in the economic benefits and *de facto* also acts to wean them away from their dependence on forests. In other words, integral to these programmes is a subtle (in some cases blatant) redefinition of indigenous peoples' relationship to the forests. The degree of incorporation of TFRK is varied, if there is any at all; the expertise is provided by the forest department, sometimes in partnership with NGOs.

Indigenous peoples' rights to the forests are progressively minimized. In some cases, indigenous peoples do receive some economic benefits (though in others non-indigenous peoples dominate the Village Protection Committees), and in the context of the takeover of resources necessary for basic survival and consequent deprivation, these economic benefits at times are seen as a steady source of survival income. This in certain cases has superficially reduced tensions and conflicts between the forest department and the indigenous peoples as a sense of co-existence is manufactured. However, this also means that the forest department is able to stifle protests and resistance. In many places, where the indigenous peoples' political consciousness is well developed, they keep away from or oppose these programmes leading to conflicts, as has happened in parts of Madhya Pradesh.

India's report to UNFF3 (2002) describes measures taken to make use of indigenous capacity and local knowledge in participatory forest management. Clearly, the picture painted by this report is not matched by the situation on the ground.

INDIAN FOREST LAW AND POLICY

After Independence, however, and according to the newly enacted Indian Constitution, forests were placed under the so-called 'state list' according to which state legislatures have a primary right to make laws. In 1976, the Indian Forest Act was added to the concurrent list of the Constitution of India, giving the Central Government and states shared responsibility and control over forest matters. Thereby the Government of India does have the power to legislate on forestry issues but only after consulting the states. The responsibility for administering the forests lies primarily with the state government. The relative position of the Government of India depends on several factors, such as its contribution to the state budget and state polities. The balance of power between central and state governments has remained a key issue in forest management ever since.

After 1947, in a post-independent India, commercial exploitation and degradation of India's forests increased dramatically. Indeed, the 1952 National Forest Policy set out guidelines which were, for the most part, directed towards the supply of cheap timber and non-timber forest products for state-sponsored industrialization and modernization. In 1980, with the passage of the Forest Conservation Act, the central government reasserted some of its control over forest-based resources. The Act restricts the state government's power to de-reserve a forest, and it restricts the use of forestland for non-forestry purposes without the prior approval of the central government. Unilateral decisions from the centre have prevailed from then onwards. Only six centres were set up to monitor forests and conservation (as per the Act), which is obviously insufficient for effective regional implementation in a country the size of India. Thus the Forest Conservation Act of 1980 has been problematic for a number of reasons, and has achieved little improvement in the 'conservation' of India's forests.

The National Forest Policy of 1988 envisaged people's involvement in the development and protection of forests *for the first time*. It stipulated that the requirements of people living in and near forests for fuel wood, fodder and small timber should be treated as the top priority, and forest communities should be motivated to identify themselves with the development and protection of forests from which they derive benefits. A primary task of all agencies responsible for forest management, including the forest development corporations, should be to engage tribal peoples closely in the protection, regeneration and development of forests, as well as to provide gainful employment to people living in and around the forests. However the 1988 National Forest Policy has never been translated into law. It remains essentially a broad statement of government intent and does little in the way of specifying any legal rights or duties owed to forest communities-especially the tribal/indigenous peoples.

CONCLUSION

Although most people know what a forest is, a definition of it which suits all cases is by no means easy to give. Manwood, in his treatise of the *Lawes of the Forest*, defines a forest as "a certain territory of woody grounds, fruitful pastures, privileged for wild beasts and fowls of forest, chase and warren, to rest and abide in, in the safe protection of the king, for his princely delight and pleasure."

This primitive definition has, in modern times, when the economic aspect of forests came more into the foreground, given place to others, so that forest may, in a general way, now be described as " an area which is for the most part set aside for the production of timber and other forest produce, or which is expected to exercise certain climatic effects, or to protect the locality against injurious influences." As far as conclusions can now be drawn, it is probable that the greater part of the dry land of the earth was, at some time, covered with forest, which consisted of a variety of trees and shrubs grouped according to climate, soil and configuration of the several localities. The conditions for an uninterrupted regeneration of the forest were favourable and the result was vigorous production by the creative powers of soil and climate.

Then came man and by degrees interfered, until in most countries of the earth the area under forest has been considerably reduced. The first decided interference was probably due to the establishment of domestic animals; men burnt the forest to obtain pasture for their flocks. Subsequently similar measures on an ever-increasing scale were employed to prepare the land for agricultural purposes. More recently enormous areas of forests were destroyed by reckless cutting and subsequent firing in the extraction of timber for economic purposes.

It will readily be understood that the distribution and character of the now remaining forests must differ enormously. Large portions of the earth are still covered with dense masses of tall trees, while others contain low scrub or grass land, or are desert. As a general rule, natural forests consist of a number of different species intermixed; but in some cases certain species, called gregarious, have succeeded in obtaining the upper hand, thus forming more or less pure forests of one species only. The number of species differs very much. In many tropical forests hundreds of species may be found on a comparatively small area, in other cases the number is limited. Burma has several thousand species of trees and shrubs, Sind has only ten species of trees.

Central Europe has about forty species and the greater part of northern Russia, Sweden and Norway contains forests consisting of about half a dozen species. Elevation above the sea acts similarly to rising latitude, but the effect is much more rapidly produced. Generally speaking, it may be said that the Tropics and adjoining parts of the earth, wherever the climate is not modified by considerable elevation, contain broad-leaved species, palms, bamboos, &c. Here most of the best and hardest timbers are found, such as teak, mahogany and ebony.

The northern countries are rich in conifers. Taking a section from Central Africa to North Europe, it will be found that south and north of the equator there is a large belt of dense hardwood forest; then comes the Sahara, then the coast of the Mediterranean with forests of cork oak; then Italy with oak, olive, chestnut, gradually giving place to ash, sycamore, beech, birch and certain species of pine; in Switzerland and Germany silver fir and spruce gain ground. Silver fir disappears in central Germany and the countries around the Baltic contain forests consisting chiefly of Scotch pine, spruce and birch, to which, in Siberia, larch must be added, while the lower parts of the ground are stocked with hornbeam, willow, alder and poplar.

In North America the distribution is as follows: Tropical vegetation is found in south Florida, while in north Florida it changes into a subtropical vegetation consisting of evergreen broad-leaved species with pines on sandy soils. On going north in the Atlantic region, the forest becomes temperate, containing deciduous broad-leaved trees and pines, until Canada is reached, where larches, spruces and firs occupy the ground. Around the great lakes on sandy soils the broad-leaved forest gives way to pines. On proceeding west from the Atlantic region the forest changes into a shrubby vegetation and this into the prairies.

Farther west, towards the Pacific coast, extensive forests are found consisting, according to latitude and elevation above the sea, of pines, larches, fir, Thujas and Tsugas. In Japan a tropical vegetation is found in the south, comprising palms, figs, ebony, mangrove and others. This is followed on proceeding north by subtropical forests containing evergreen oaks, *Podocarpus,* tree-ferns and, at higher elevations, *Cryptomeria* and *Chamaecyparis.* Then follow deciduous broad-leaved forests and finally firs, spruces and larches. In India the character of the forests is governed chiefly by rainfall and elevation.

Where the former is heavy evergreen forests of Guttiferae, Dipterocarpeae, Leguminosae, Euphorbias, figs, palms, ferns, bamboos and india-rubber trees are found. Under a less copious rainfall deciduous forests appear, containing teak and sal (*Shorea robusta*) and a great variety of other valuable trees. Under a still smaller rainfall the vegetation becomes sparse, containing acacias, *Dalbergia sissoo* and Tamarix.

Where the rainfall is very light or *nil,* desert appears. In the Himalayas, subtropical to arctic conditions are found, the forests containing, according to elevation, pines, firs, deodars, oaks, chestnuts, magnolias, laurels, rhododendrons and bamboos.

Australia, again, has its own particular flora of eucalypts, of which some two hundred species have been distinguished, as well as wattles. Some of the eucalypts attain an enormous height.

2

Forest Soil

FOREST SOILS AND CARBON STORAGE

Introduction

The global cycling of C involves fluxes between the fossil C reservoir, the atmosphere, the oceans and the terrestrial biosphere. The role of the terrestrial biosphere in buffering or contributing to increases in CO_2 of the global C cycle is uncertain, partly because of the difficulty in predicting the likely impact of climate change, associated with rising atmospheric CO_2, on C flux. Because forests and wooded lands cover about one-third (4100 million ha) of the land area of the earth, they are a significant component of that C stored in the terrestrial biosphere which is exchangeable with the atmospheric pool of C. The estimated 787 Pg of C in forest soils and associated peats is more than two-thirds of the total of 1146Pg of C held in forests.

About 33% of forest soils are covered by coniferous forests in the high latitudes of Russia, Canada and Alaska. About 25% of forest soils are covered by forests in the mid-latitudes of continental USA, Europe and Australia, while the low-latitude (tropical) forests of Asia, Africa and the Americas account for 42% of forested land.

More than half of the low-latitude forests are in tropical America. The amount of C stored in forest soils increases with northern latitude, so that low-latitude (tropical) forests contain about 27% of the pool of soil C, increasing to about 60% in soils of high-latitude northern boreal forests. Thus the soils of high-latitude northern forests are an important C pool and any change in the climate or the management of these forests that diminishes the soil C pool will significantly affect global C storage. This applies equally to the tropical forests of lower latitudes, where soil C stocks are less than half those of the boreal forests, but where small increases in temperature will have a large proportional effect on metabolic activity.

The balance between net production or consumption of CO_2 in forests is largely determined by decomposition rate and the rate of soil organic C formation.

This balance can be estimated reasonably well for undisturbed forests. However, following clearing of forests the relatively low-level sinks of CO_2 in forest soils become high-level sources. Conversely, reforestation may restore the sink capacity of the soil, thereby off-setting to some extent the increase in atmospheric CO_2 due to land clearing and the oxidation of fossil fuels. Extensive tree planting and the development of policies for 'carbon credits' has therefore focused research efforts on the amount and stability of soil C and on processes that control the inputs and outputs of C.

Because soils contain the largest pool of C in terrestrial ecosystems, small changes in soil temperature or moisture or both could lead to a substantial increase in atmospheric CO_2 concentrations. Current general circulation models predict that the temperature of the earth's atmosphere will increase by 1°C over the next 35 years, altering global climate patterns and hence impacting on many ecosystem processes. Climate warming is predicted to be greatest nearest the poles so that boreal and tundra biomes will be influenced the most; this has generated considerable interest in the interaction of climate with the biogeochemistry of these biomes.

Carbon accumulates in forest soil as a result of litterfall and root input, while outputs result from microbial degradation of organic matter, eluviation, solution losses and erosion. The carrying capacity for soil C increases as forests mature, and the maximum carrying capacity is controlled by climate, topography, soil type and vegetation. For example, Niklinska *et al.* (1999) measured the effect of increasing temperature on the rate of respiration of humus from forests of Scots pine and predicted an increase in the annual rate of soil respiration for the high latitudes of the Northern Hemisphere of 70 Tg ($1Tg=10^{12}g$) CO_2 with a 2°C increase in temperature. Niklinska *et al.* (1999) therefore suggested that northern forests may become a net C source with increased global temperatures. Much of the global store of soil C is sequestered in peats and peaty soils developed in cold and wet climates. Drying of these organic soils as a result of climate change will significantly increase the rate of loss of soil C, although the magnitude of change of C-flux following the drainage of peatlands and their conversion to conifer plantations has been difficult to measure.

In general, the state of current knowledge is insufficient to predict whether rising temperatures will lead to a net release or storage of C in the terrestrial biosphere. The potential for forest soils to store or release C under changing climate and land use has been modelled on a large scale by Schimel *et al.* (1994).

The CENTURY model of the turnover of soil organic matter applied to global forest and grassland ecosystems showed that soil texture and lignin concentration of foliage are key variables influencing soil C storage. Feedback between the N cycle and NPP influences model predictions of the terrestrial C balance. All projections, based on a range of ecosystem models, suggest that increasing CO_2 will increase C inputs to soil, which will in turn widen the detrital C: N ratio, thereby decreasing the rate of decomposition.

Simulations using the CENTURY model show that N cycles more rapidly under warmer temperatures and that as soil organic matter is lost more N becomes available for plant growth, resulting in more soil organic matter formation, and thereby providing negative feedback. It seems that the efflux of organic C from soil with global warming would be much greater, perhaps double the current efflux, without this so-called 'negative feedback'. However, speculation on the role of increased N mineralization from soil organic matter in increasing the vegetation biomass has not been substantiated with field data.

The Intergovernmental Panel Assessment on Climate Change (IPCC) assessed the role of forests in the terrestrial C cycle over the decade 1980-90. The assessment concluded that regrowth forests of the Northern Hemisphere may have been a net sink of 0.5Gtyear^{-1} of C, while deforestation in the tropics released 1.6Gtyear^{-1} of C. Climate change in tropical regions also has the potential to alter C storage significantly, as shown by Tian *et al.* (1998) in a study of the effect of interannual climate variability on C storage in Amazonian ecosystems.

The hot dry weather of El Niño years in Amazonia decreased soil moisture, leading to a net release of 0.2Pg of C in 1987 and 1992. In other years, the forests and savannahs of the Amazon basin are a sink of up to 0.7Pg annually of C. These findings are supported by Silver (1998) who predicted that small increases in temperature and CO_2 will result in more rapid decomposition and therefore an increased CO_2 flux from moist tropical soils.

Deep soils of tropical forests contain a significant proportion of the ecosystem C, with estimates of the mass of C at depths below 1 m equal to or greater than the mass of C in the above-ground forest. This C, held deep in the soil, represents a significant source of CO_2 emissions following deforestation and conversion to pasture.

The impacts of forest regrowth, of increasing atmospheric concentrations of CO_2, of N deposition and of global warming on C storage in forests remain critical areas for future research. Holland *et al.* (1997) modelled the effects of atmospheric NH_X-N and NO -N deposition on global terrestrial C and estimated an increased rate of storage of C of 1.5-2.0Gtyear^{-1}, an amount similar to that of the missing terrestrial sink. Makipaa (1995) found increases in C storage in both humus (14-87%) and mineral soil (15-167%) following long-term N addition experiments in boreal forest. In temperate forests, N deposition is expected to both alleviate N deficiency and increase C sequestration in biomass and soil organic matter.

Houghton (1996) claimed that the predicted increase in mid-latitude forest sequestration of C has occurred (data from Canada, USA, Europe and the former USSR), with a greater than expected accumulation of C (0.8Pgyear^{-}

^{1}) in biomass and soil pools during the 1980s. These gains of C in forests of the mid-latitudes have been offset by deforestation in the tropics, which contributed 1.6Pgyear^{-1} of C during the 1980s.

Forest management to conserve and sequester C has been identified as one of three main strategies to stabilize greenhouse gases. Johnson *et al.* (1995) likewise identified significant global-scale opportunities for both reducing C emissions from soil and increasing C sequestration in soil. The viability of such strategies has already been questioned on both sociological and scientific grounds. Management to increase C storage in forests has been investigated at regional, national and global levels. In a global survey of 94 nations, Dixon *et al.* (1994) found that the establishment of agroforestry and alternative land-use systems (fuelwood and fibre plantations, bioreserves, intercropping systems, shelterbelts and windbreaks) on marginal and degraded lands could sequester 0.82-2.2Pg annually. The estimated cost of these activities was $US1–69 per tonne of C, which compared favourably with other options.

Deforestation, especially of low-latitude forests, remains one of the main causes of the global net release of CO_2 from forest soils to the atmosphere, yet the magnitude of emissions is uncertain due to limited studies of changes in soil C following forest clearing. Dixon estimated that 15.4 x 10^9ha of forest were lost in the two decades between 1971 and 1990; the release of C from these soils will continue for decades as the C content of the soil gradually stabilizes following the change in land use.

Torn *et al.* (1997) correlated the content of soil organic C with soil age in a forested chronose-quence in Hawaii and concluded that the accumulation and subsequent loss of organic matter was largely driven by changes in the cycling of mineral-stabilized C over millennia rather than by changes in either the amount of fast-cycling organic matter or NPP Soil mineralogy is therefore important in determining the quantity of organic C stored in soil, its turnover time, and C exchange between ecosystems and the atmosphere during long-term soil development. The C storage capacity of the 25% of the world's mineral soils that began developing after the last major glaciation will change as they undergo further development, and they may eventually become long-term sources of atmospheric CO_2.

Atmospheric Concentrations of Carbon Dioxide

Increases in atmospheric concentrations of CO_2 are probably the primary driving force of the anthropogenic greenhouse effect and thus of climate change; however, the atmospheric concentration of CO_2 also directly affects plant growth. Atmospheric CO_2 is the source of the plant's carbon and its active uptake in photosynthesis depends on diffusive flux across a concentration gradient. Trees grow better in elevated CO_2 compared with preindustrial ambient values (=240 ppm) when other environmental factors, particularly nutrient and water availability, are not limiting. However, when these factors are below optimum, trees suffer from increased stress in elevated CO_2. Furthermore, CO_2 concentrations directly affect many plant processes other than photosynthesis, including respiration, stom-atal conductance, ratio of root to shoot mass, rates of leaf emergence, leaf area index (ratio of leaf area to ground area), growth, senescence and tree phenology. The potential for the effects of elevated CO_2 concentrations to interact with those of other factors is considerable. Experimental evidence from controlled exposures is now showing that such interactions occur and can result in enhanced susceptibility to a number of abiotic stresses, particularly drought, nutrient deficiency, frost damage and damage from other air pollutants.

Although the growth responses of trees to elevated CO_2 are more often positive than negative, they vary greatly with species and the assumption that the overall effects of the current rise in CO_2 concentrations will be beneficial to forestry is an oversimplification that cannot be justified by detailed consideration of the evidence. The data reported by Broadmeadow and Freer-Smith (1999) are representative of similar interactions reported in the literature. Oak *(Quercus petraea)* showed a positive growth response to twice current ambient CO_2 concentrations (700 ppm); moreover, the detrimental effects of realistic peaks of ozone (O_3), *i.e.* a diurnal cycle peaking at 80ppb, were lost when *Q. petraea* was grown in 700 ppm CO_2. Two other European species (ash, *Fraxinus excelsior,* and Scots pine, *Pinus sylvestris)* showed no detrimental effects of O_3 exposure and smaller stimulations of growth in 700 ppm CO_2. As has been reported for most trees exposed to realistic doses of these pollutants, the stimulation of growth in oak was the result of increased photosynthesis, while the chronic growth reduction due to O_3 resulted from reduced photosynthetic capacity.

A perhaps unique feature of this particular experiment was that soil-rooted (rather than pot-rooted) plants were also grown with two different soil water regimes in all combinations of CO_2 and O_3 (droughted and irrigated). The detrimental effect of O_3 on photosynthesis was reversed by elevated CO_2 but only for plants experiencing drought. Protection by CO_2 from O_3 probably results from the increased internal leaf CO_2 concentrations, closing stomata and limiting O_3 uptake ; a classic stress avoidance mechanism in Levitt terminology. In well-watered trees, elevated CO_2 failed to decrease stomatal conductance and thus failed to provide protection from O_3. Unfortunately, tree responses to CO_2 and the interacting effects of different abiotic factors are even more complicated in forest stands.

A comprehensive consideration of the biochemical and physiological effects of CO_2 is beyond the scope of this chapter. This overview is based on the recent compilation of results on elevated CO_2 effects on trees resulting from the investigation by the European Union (EU) ECOCRAFT project. Within this programme, comprehensive studies of eight or more different tree species were conducted using a number of different techniques. The key processes affected were stomatal conductance (both stomatal opening and stomatal density), photosynthesis (ribulose 1,5-bisphosphate carboxylase oxygenase (Rubisco) content and activity, chlorophyll content and the light reactions), respiration (although CO_2 effects seem variable and small), phenology (particularly where nutrient limitations towards the end of the growing season give early bud set and senescence) and allocation.

Compiled as a single volume, the studies conducted within ECOCRAFT present a coherent picture of increased growth in elevated CO_2 and of the very major way in which nutrient availability and water deficit can modify this response. The growth increases in elevated CO_2 in all the experiments conducted varied between 5% (Sitka spruce) and 58% (beech). In forest ecosystems water and nutrient availability are closely linked and experiments have also been conducted on the three-way interactions of elevated CO_2 concentrations, nutrient availability and drought on tree growth. These interactions are of major significance in determining stress response and forest condition.

Nitrogen Depositions

Nitrogen is important as a plant nutrient because it occurs in many plant proteins (particularly Rubisco, the enzyme that catalyses the first carbon fixation reaction of photosynthesis) and chlorophyll. Although N_2 makes up about 78% of the atmosphere by volume, this is not available to plants. Plants absorb almost all N from the soil as nitrate (NO_3~) or, in acid soils, as ammonium (NH_4^+). First, atmospheric N_2 has to be fixed (reduced) to organic nitrogen by soil microorganisms or in the roots of N-fixing plants (*e.g.* alders, *Acacia*). Most of the N in soils is in the form of organic N, which is unavailable to tree roots. Organic N is then converted to NH_4^+ by ammonification, which in warm near-neutral soils is rapidly oxidized to NO_3~ (nitrification) by soil bacteria that derive energy from the conversion. In acid forest soils, nitrifying bacteria are less abundant and NH_4^+ becomes a more important source of N. Low temperatures in some forest soils will also lower NO_3~ availability. Ammonification consumes protons but nitrification generates protons. The net result of the conversion of organic material to NH_4^+ and then NO_3~ is acidification. As a cation, NH_4^+ can be held on cation exchange sites, while NO_3~ is more likely to be leached from soils. Nitrogen is also lost from soils by denitrification, the production of N_2, NO, N_2O and NO_2 from NO_3~ by anaerobic bacteria, often deep in soils or in waterlogged or compact soils.

Small amounts of N are also passed from the atmosphere to soil as NO_3~ and NH_4^+ in rain (up to 30 kg ha^{-1} annually) and these ions can then be absorbed by plants. Natural sources of atmospheric NO_3~ and NH_4^+ are forest fires, volcanic activity and the oceans and NO_3~ production in lightning; however, industrial processes and the agricultural use of nitrogenous fertilizers and intensive animal husbandry are major anthroresult from large NH_4^+ inputs. This will inhibit base cation uptake resulting in K or Mg deficiency. Soil acidification from NO_3~ or NH_4^+ inputs will also reduce the decomposition of organic matter so that it can accumulate in soils; litter accumulation.

There is a limit to the amount of N that a forest ecosystem can use and this limit is likely to be set by the availability of other nutrients, water, light or other environmental factors. A forest is said to be N saturated when it cannot retain any additional N and leaching occurs. The concept of N saturation has become central to discussion of forest soil critical loads.

Forms and Turnover of Carbon in Soil

Organic matter includes most of the litter and humic layer in forest soils and generally forms 1-10% of surface mineral horizons. The comminution and decomposition of above- and below-ground litter in forests releases CO_2, and nutrient elements are mineralized and immobilized. While all organic matter is ultimately oxidized in forest soils, compounds that decompose over centuries and millennia are synthesized in the process of humification, which involves processes of both degradation and synthesis. This humus is amorphous and is intimately mixed with soil mineral components; it is difficult to describe in any precise chemical manner. The inert organic matter of forest soils is protected from decomposition in a number of ways, including association with mineral particles, which is the basis for longer turnover times for organic matter in clay-textured soils relative to sandy soils. Most of the organic matter in soils is very old, and decomposition to inorganic constituents will take centuries to millennia.

Both chemical and physical methods have been used to fractionate soil organic matter; however, it has not been possible to separate organic matter into discrete components of varying decomposability. Traditional chemical methods date from the classical empirical fractionation based on sequential acid and alkali extractions; this recognizes fulvic acids (soluble in both acid and alkali), humic acids (soluble in alkali and precipitated by acid) and humin (insoluble in both alkali and acid). The structure of humic compounds is not accurately known but humus formation involves the polymerization of aromatic compounds, many of which are derived from lignin and polyphenolic pigments.

The application of analytical techniques such as nuclear magnetic resonance, Fourier transform infrared spectroscopy and pyrolysis-mass spectroscopy are refining our understanding of the structure and reactiveness of humic compounds in soil. Recent studies of humic compounds in forest soils have shown that the stability of the inert or occluded organic matter appears to be inversely related to the content of *O*-alkyl C and directly related to the content of aromatic C. Physical methods based on density usually separate soil according to free particulate organic matter (macro-organic matter), occluded particulate organic matter (light fraction) and colloidal or clay-associated organic matter (heavy or humified fraction). Density fractionation, based on dispersal of soil in NaI solution, has proved useful in separating organic matter into components that are physically and chemically distinct, and which are mineralized at different rates by soil microorganisms. Generally the light fraction extracted in NaI comprises partially decomposed roots and litter and is relatively labile; it has been reported to represent 25-50% of total soil carbon in forests of the United States.

By contrast, the heavy fraction is regarded as more resistant humic material that is adsorbed on to mineral surfaces. The heavy fraction can be further separated into proximate fractions such as soluble fats and oils, celluloses and lignin, and phenolic compounds. Studies employing these density-fractionation techniques have established that old-growth forests store more C than young-growth forests and that soils of old-growth forest contain greater percentages of recalcitrant organic matter typically richer in lignin and tannic compounds.

Further application of these methods, which identify the most biologically active forms of C in forest soils, are critical to developing a better understanding of soil organic matter turnover under a range of influences.

Turnover time ($1/k$) is defined as the time that C is held in organic forms per unit volume of forest floor and soil; it is therefore the time taken for all of the organic C in the soil to be replaced. Turnover times range from hours and days for freshly added and easily decomposed organic matter to centuries and millennia for highly stabilized organo-mineral complexes in the soil.

As apparent from the discussion of C storage in forest soils, much of the recent interest in forms of organic matter in forest soils and their turnover time is in relation to the impacts of climate change and forest management on pools and fluxes of C. Models of soil C turnover have been used to predict the effects of climate change and of different land uses on soil C fluxes. While a number of models have been developed, the most popular are those based on the CENTURY model and the ROTH-26.3 model. These models have been used to simulate changes in organic C of arable and grassland soils with accuracy.

However, studies in forest soils must encompass organic matter with a wide range of decomposition constants. Although the C models were originally developed for agricultural soils, they have recently been adapted to model the more complex environment of forest soils by inclusion of modules to accommodate the forest floor and litter layer. An adapted CENTURY model predicted the pattern or ranking of decomposition of surface litter in a matrix of *Metrosideros* sites on Mauna Loa but did not predict the magnitude of differences among sites. The refinement of the CENTURY model as it applies to forest soils will doubtless demand a better understanding of the reactivity of organic matter on the forest floor. Studies of the proportion of the large store of soil C that exchanges with the atmosphere within centennial and shorter time-scales (compared with time-scales of millennia) are critical to our understanding of the global C cycle. At the scale of landscapes and over geological time, Torn *et al.* (1997) demonstrated that the turnover time for the large, passive pool of soil C is controlled primarily by soil mineralogy.

Forest Soils and Acidic Inputs

Glaciers have advanced and retreated across Massachusetts many times over the past several million years. The last glacier retreated over 10,000 years ago. Our soils are the product of the interaction of vegetation and climate on what the glaciers left behind. Most upland, forested soils have developed out of what is called `glacial till,` that is, soils that were created and mixed by glaciers moving across the landscape. They are typically stoney, fine-textured soils on top of bedrock.

They may or may not have hardpan layers. We also have pockets of `glacial deposit` soils which were left by streams and rivers created by melting glaciers. These deposit soils are made of mixtures of sand and gravel and are found where there were edges of tongues of the retreating glaciers--mostly in river and stream valleys. Many hillsides have scattered pockets and bands of these soils among the predominantly glacial till soils. There are also lacustrine and alluvial soils. These are both silty soils, the former from deposits on old lake bottoms and the latter from flood deposits along rivers and streams. Most of these soils are in agricultural use.

Sulphur emissions have increased since the start of the Industrial Revolution to the extent that 75% of global emissions of S (total about 105Mtyear^{-1}) come from anthropogenic sources. While emissions from natural sources are about evenly distributed between Northern and Southern hemispheres, 90% of anthropogenic sulphur emissions come from sources in the Northern Hemisphere (about 70Mtyear–^{1} in the Northern Hemisphere and 7.7Mtyear^{-1} in the Southern Hemisphere.

Forest Decline

Forest decline in the Northern Hemisphere was increasingly observed during the 1970s and 1980s and was initially linked with the increasing acidity of rain, the latter caused by nitric and sulphuric acid formed from emissions from industry, cars, etc. However, there are other sources of acidity such as dry deposition (including both direct gaseous absorption and particulate deposition), and the rates of dry deposition of SO_2 and NH_4^+ may exceed rates of deposition in rainfall. There are also many causes of forest decline other than acid deposition. There is little doubt that acid rain (used here in a generic sense) is a cause of some severe forest declines, particularly when close to emission sources. The most extensive studies have been in Germany and the results can be simply summarized.

With increasing acidity, cation exchange capacity is reduced (leading to the loss of Ca^{2+} and Mg^{2+}) and the concentration of Al ions increases. Ca : Al and Mg : Al ratios in needles of *Abies alba* from healthy stands are greater than those in needles from stands that show symptoms of forest decline.

Lindberg *et al.* (1986) estimated that atmospheric deposition supplied 40% of the annual requirement for wood production of S and 100% of that for N. Continuing input of acid rain therefore produces imbalances in the ratio of N to cations in tree crowns that result in chlorosis and, if serious enough, death. However, for many forests cause and effect are not easy to link. Most forest soils are acidic, so there is the overriding problem of assessing increasing acidification of an already acidic medium; added to this is the difficulty of determining and assessing the sources of acid input.

Two themes show the complexity of the acid rain problem. Markewitz *et al.* (1998) analysed soil samples from the Calhoun Experimental Forest that had been archived over three decades. They concluded that natural processes of acidification (increase in biomass and uptake of $[\text{cations}]_c > [\text{anions}]_c$ by plants, loss of $[\text{cations}]_c > [\text{anions}]_c$ by leaching, increase in soil organic matter, mineralization of N to NO_3^-, microbial and root respiration, production of organic acids) accounted for 68% of acidification over the 30 years, with acid deposition accounting for the remaining 32%. With a threefold increase in SO_2 emissions for the eastern USA from 1960 to 1980, the capacity of the surface soil to adsorb SO_4^{2-} was saturated but there was strong adsorption below 60 cm. A result of major interest is that, from 1982 to 1990, the concentration of extractable SO_4^{2-} in soils at 35-60 cm decreased significantly. This decrease was equivalent to a loss of charge of 2.2kmol$_c$ha^{-1}year^{-1}, and Markewitz *et al.* (1998) suggest that the decrease in concentration of extractable SO_4^{2-} shows that the soils at Calhoun are responding (or recovering) to the decrease since 1980 in rates of both SO_2 emission and SO_4^{2-} deposition.

For the intensively studied Hubbard Brook Experimental Forest, north-eastern USA, inputs of acid rain have caused large losses of cations (particularly Ca^{2+} and Mg^{2+}, together with K^+ and Na^+) in drainage waters. Since 1963, atmospheric emissions have decreased, resulting in decreases in concentrations of SO_4^{2-} and the cations Ca^{2+}, Mg^{2+}, Na^+ and K^+ in bulk precipitation.

The loss of cations in drainage and the net uptake of cations by the forest together represent an 'ecosystem requirement' that must be balanced by inputs in bulk precipitation, from weathering and from release of cations from exchange sites in the soil. For Ca^{2+}, precipitation supplied 29% and net release from soil 63% of this required balance during the period 1940-55. From 1976 to 1993, precipitation supplied only 12% of demand and net release from soil increased to 79%. Likens *et al.* (1996) therefore concluded that even with major reductions in emissions, the response of soil and streamwater chemistry (particularly the acid-base balance) will be significantly delayed relative to that expected from studies of inputs of acids (the bio-geochemistry of sulphur) alone.

Nitrogen Saturation

The second theme is called N saturation, the hypothesis that forests may become 'N saturated' as a result of too much N from acid deposition. Again, this is a concern for the Northern Hemisphere forests, for example in Europe , in North America and in Japan. Binkley and Högberg (1997) examined the hypothesis of N saturation in considerable analytical detail for forests in Sweden. While S deposition rates in Sweden have decreased by as much as 50% since 1980, the annual rates are still as high as 25 kg ha^{-1} along the south-west coast, decreasing to <12kgha^{-1} in the north. These authors also estimated the annual rates of input of N to the south-west coast as 8-12 kg ha^{-1} for NO_3–-N and 5-15 kg ha^{-1} for NH_4^+-N, decreasing in the north of Sweden to <2kgha^{-1} for both NO_3–-N and NH_4^+-N.

Despite these sometimes high inputs of N, the N is apparently accumulated since in only a few stands near the south-west coast do losses of N by leaching approach inputs of N in deposition. Thus the evidence is that Swedish forests are not N saturated, and forest decline is not linked with any condition of N saturation. The evidence from the application of N fertilizers is uncertain to say the least; some stands respond positively, some negatively and some not at all. Nevertheless, Binkley and Högberg (1997) concluded that 'the vast majority of conifer forests in Sweden still suffer from chronic N deficiency'.

Because of concerns about the acidification of forest soils a number of countries have initiated liming experiments, such as the Höglwald experiment in Germany, where 4tha–^{1} of dolomitic limestone was applied.

In reviewing these experiments, Binkley and Högberg (1997) concluded that 'the overall effect of liming of forest soils in the Nordic countries has been a general decrease in forest growth (of perhaps 10%)'. A piece of jargon that has emerged from soil acidification and the N-saturation hypothesis is 'vitality fertilization' (or, even worse, 'vitalization fertilization'), a term that really means no more than the application of a fertilizer to ameliorate a deficiency (or, more ornately, to correct a nutrient imbalance). We agree strongly with Binkley and Högberg (1997): 'vitality fertilization' should not be used because it implies lack of vitality where, as in most forests, there is no symptom of decline.

Acid deposition is a problem for the Northern Hemisphere, simply as a consequence of the very much greater density of land area and population in the Northern than in the Southern Hemisphere. While forest decline is caused by acid rain in many forests, in many other forests in both hemispheres decline cannot be linked with high inputs of acid. Wild (1993) summarized a number of causes of decline other than acid deposition.

- Natural acidification processes lead to accumulation of Al and a deficiency of Mg in old trees. This problem is exacerbated by an increase in N supply that causes the development of large crowns, placing demands on Mg. Where Mg is in short supply, Mg is translated from older foliage to developing foliage.
- The increased input of N produces softer or more palatable tissues, resulting in increased insect attack.
- Ozone is directly absorbed by leaves and is extremely damaging to structure and function; decline may therefore be caused by increased inputs of ozone.
- The 1970s and 1980s in Europe were characterized by hot dry summers and cold winters; these conditions might place stress on some species in some areas.
- Poor forest management.

In summary, the 1980s saw the widespread prediction of a general decline of European forests (the concept of *Waldsterben).* By the 1990s this view was greatly modified, particularly by evidence that the growing stock of European forests increased between 1971 and 1990 by 25% and that growth of the forests increased by 30%. Skelly and Innes (1994) concluded that 'the concept of a general forest decline is untenable', and that the evidence for air pollution causing forest decline 'away from known sources of pollution is extremely limited'.

Carbon Dioxide Exchange

The Policy Imperative

The exchange of CO_2 between forests and the atmosphere has become a topic of intense interest over the past 3 years as the role of forests in the global atmospheric CO_2 budget has become more widely appreciated. The Kyoto Protocol, arising from the Third Conference of the Parties (COP3) of the United Nations Framework Convention on Climate Change (UNFCCC) held in Kyoto in December 1997, explicitly implicated forests as agents in the net removal of CO_2 of anthropogenic origin from the atmosphere. The developed countries of the world were assigned legally binding targets for the reduction of their industrial emissions of CO_2 to the atmosphere, a part of which could be offset against the net uptake of CO_2 by new, young forests. The Intergovernmental Panel on Climate Change (IPCC) was commissioned to produce a special report on 'Land Use, Land Use Change and Forestry' for submission to the UNFCCC at COP6 in November 2000 and there is an expectation that 'full accounting' of all forests may then be included. The overall result of these developments has been a major upsurge of interest and concern regarding the exchanges of CO_2 between forests and the atmosphere on stand, forest, national, regional and global scales, as is happening now and is likely to eventuate over the next 100 years.

Forests as a Carbon Pump

Forests are the major terrestrial interface for C transfer between the atmosphere and the soil. Like the net exchange of short- and long-wave radiation and the balance between precipitation and evaporation, the NEE of CO_2 between forests and the atmosphere is a balance between two largely simultaneous, co-located, opposed sets of processes.

In all three cases it is the balance between the inflows and outflows that is of major consequence. We are concerned with the small shift in the radiation balance responsible for global warming, with changes in the spatial and temporal balance between precipitation and evaporation that may lead to water shortage and, in forest stands, with the size and permanence of the balance between the inflows of C in photosynthesis by the trees, understorey and ground vegetation and outflows resulting very largely from the heterotrophic respiration in the soil and from the corresponding autotrophic respiration of the trees.

We may regard the trees in forests as 'pumps' that effect a net transfer of C from the atmosphere to the soil where it is sequestered. The large amounts of C that have accumulated in forest soils in the Northern Hemisphere since the retreat of the ice sheets after the last glaciation are evidence of this. While the trees in a forest are juvenile and growing rapidly, they may also accumulate significant amounts of C in the wood; however, even when the trees 'mature' and are growing very slowly, if at all, they none the less continue to transfer C from the atmosphere to the soil where it accumulates. This transfer is effected through leaf and branch fall, internal translocation of assimilate to the root system and fine-root turnover.

For example, trees in large areas of the boreal forest today are small in diameter (*e.g.* stands of black spruce with a median diameter of 10 cm at 120 years old) and are growing on deep peat-rich soils, the C of which has all accumulated over the past 8000 years or so. On the basis of the 'climax concept', it was long believed that pristine forests, particularly tropical forests, were in a state of equilibrium such that over a period of several years they were C neutral, with neither a net gain nor loss of C. This view has been challenged in recent years with increasing evidence from eddy covariance measurements of NEE of a net gain in the exchange of CO_2 and from sample plot studies of NPP which show that tree growth in undisturbed tropical moist forests continues to sequester C. Regarding forests as a pump transferring C from the atmosphere to the soil in this way is closely analogous to the 'biotic pump' in the photic zone of the oceans, which transfers C from the atmosphere to long-term sequestration in the deep ocean.

Dynamics of Carbon Dioxide Exchange

Daily Dynamics

Over the course of a day, C sequestration essentially follows solar radiation (provided there is no major constraint such as frozen or dry soil), with C being accumulated during the daylight hours and lost at night. For example, large daytime gains by tropical forest stands are offset by large night-time losses. Depending on the balance between these hourly gains and losses, for any stand there may be a net gain or loss of C over any 24 h in the year.

Typically, through a growing season, there may be a net gain of C on about three-quarters of the days but a loss on days of particularly low solar radiation input (*i.e.* with cloud and rain, or smoke in the atmosphere) and high temperatures.

Seasonal Dynamics

Over a year, the length of the photosynthetic season has a major influence on C transfer. In boreal coniferous forest, for example, one-third to a half of the C gained in the five summer months is lost by respiration over the seven winter months, during which a small efflux of C continues although the ground is frozen.

In broadleaved temperate forest the length of the photosynthetic season is determined by budburst in the spring and leaf senescence in the autumn, whereas in temperate coniferous forest the season of net photosynthetic gain is determined by daylength and daily total radiation input in the winter months and may be 10 months long for coniferous forest in maritime climates. In boreal, temperate and Mediterranean climates the vegetation is strongly seasonal and NEE may be strongly constrained by water availability for a large part of the time; only in tropical moist forests is net C gain near continuous throughout the year, being reduced by occasional short periods of low temperature and moderate seasonal water shortage.

Annual Dynamics

Transfer of CO_2 to young forest stands regenerating after disturbance such as fire or harvesting is critically dependent on the phase of the successional cycle. During the first 5 or 10 years the disturbed area of young trees is likely be losing C to the atmosphere; subsequently, as the trees come to occupy the site fully, the area will accumulate C strongly during a phase of rapid growth that may last for decades, depending on the species of trees and site conditions.

Mature forests take up C from the atmosphere at much slower rates, but even as the growth increment of the trees declines towards zero, C continues to be transferred from the atmosphere to the soil via the trees in the form of above- and below-ground detritus. None the less, old-growth forests do become net sources of C to the atmosphere under some circumstances, for example if the forest soil becomes warmer or more aerobic, thus promoting oxidation of the soil organic matter.

Net Transfer of Carbon Dioxide from the Atmosphere to Forests

Tropical Forests

So far there are few C flux measurements over periods long enough to give annual estimates of NEE in tropical forests, although they are likely to increase appreciably over the next 5 years with the onset of the LBA experiment. Eddy covariance measurements indicate an annual net C sink of about 1.0 Mg ha^{-1} in pristine, sparse, dry, rain forests in Amazonia and about 5.5 Mg ha^{-1} for dense, moist, rain forest.

Temperate Forests

Forests in the temperate region are virtually all managed to a greater or lesser extent and there are but few patches that might be regarded as pristine. Recent measurements of NEE by eddy covariance over a year or more in Europe in the EUROFLUX experiment and in North America indicate rates of C sequestration in the range 1.5 to 7 Mg ha^{-1} over about 20 sites differing in latitude, altitude, climate (continental vs. maritime), species, sparseness of cover and degree of management.

The values at the lower end of the range derive from sparse sites where lack of water is a major factor or from northern sites where length of growing season, lack of nutrients and low temperatures are major factors. Estimates of NPP of sample plots indicate annual carbon sequestration of approximately 3 Mg ha^{-1} (range 1 to 8 Mg ha^{-1}) based on numerous yield tables. A preliminary scaling-up exercise of the EUROFLUX measurements suggests that the largely temperate forests of the European Union are a current annual C sink of 0.17 to 0.35 Gt.

Boreal Forests

The extensive boreal forests across Siberia and Canada are fire-driven climax forests, the near end-point of succession following retreat of the ice after the last glaciation about 10 000 years ago. These forests have been subject to increasing exploitation over the past 150 years but much of the area standing can be regarded as largely pristine. Measurements of C sequestration by eddy covariance on a campaign basis in Siberia, over periods of several months up to 5 years in northern Canada, in the BOREAS experiment and in northern Europe have demonstrated old-growth spruce stands that are C neutral in Canada and a C source in Sweden, losing C at an annual rate of up to 1 Mg ha^{-1}. However, the majority of boreal forest stands investigated, some 12 or more, are C sinks sequestering C at annual rates of up to 2.5 Mg ha^{-1}. Estimates of NPP from sample plots also indicate that boreal forests are weak C sinks of about 2.5 Mg ha^{-1} (range 1 to 6 Mg ha^{-1}).

Carbon Balance Components

Carbon Stocks and Flows in Forest Biomes

To understand the magnitude of the net transfer of CO_2 between stands and the atmosphere at a particular stage in a succession and to appreciate its sensitivity to current conditions, disturbance and climate change, it is helpful to consider the principal component processes within the stand that add up to give the overall net exchange.

The feedbacks among these processes, their speed of response and their sensitivity to environmental change determine the present and future flows of CO_2 and the C sequestration capacity of the stand. It is evident that the annual NEE is the difference between a very large gain term (*i.e.* photosynthesis of the foliage) and a very large loss term dominated by respiration of heterotrophs in the soil. Thus the net exchange is particularly sensitive to relatively small proportional changes in the opposed gain and loss terms, particularly if they are out of phase.

A mass balance of the component fluxes above ground (photosynthesis; foliage, branch and stem respiration; leaf, branch and stem litter of all kinds, *i.e.* detritus; and above-ground NPP) enables an estimate to be made of the amount of C internally translocated below ground. A mass balance of the component fluxes below ground, comprising the inputs from above (*i.e.* the detritus and translocate), root and heterotrophic respiration, fine-root turnover, and mycorrhizal and root system NPP, enables an approximate estimate to be made of net changes in the pool of soil C.

In principle, summation of the component fluxes over a year should add up to the independently measured annual NEE. However, there are appreciable errors in measuring the below-ground components in particular, so that close agreement between measurements or estimates of the components of the budget is not as yet to be expected.

Exchanges of Volatile Organic Compounds and Methane

The net exchange of C between forests and the atmosphere also includes the emission of bio-genic volatile organic C compounds (VOC) such as isoprene and terpenes from the tree canopy. Simple algorithms have been developed to calculate emissions of individual compounds from individual tree species based mainly on PAR and leaf temperature. The emissions of VOC increase with temperature and PPFD and it has been argued that the emissions of certain VOC compounds, notably isoprene, reduce the physiological effects of water stress. The absolute magnitude of C exchange represents a small fraction of the total. Emissions of terpenes from Norway spruce in Denmark amount to no more than 0.3% of the annual C assimilated, but short-term losses from the Mediterranean species *Quercus ilex* may exceed 6% at high temperatures Globally emissions are small, around 0.1% of annual carbon emissions from vegetation.

Undrained peatlands in high latitudes have accumulated appreciable amounts of C from the atmosphere since the retreat of the ice and are currently significant annual CO_2 sinks (0.2-0.5Mgha^{-1} of C) but are also small annual sources of methane (CH_4) (0.03-0.3 Mgha^{-1} of C). By contrast, drained peatlands, whether intended for agriculture or afforestation, release C as CO_2 because of accelerated decomposition of the aerobic peat but no longer release CH_4 in significant amounts.

Peatlands drained for agriculture continue to be a sustained C source while the peat remains. Depending on species of tree and time in the rotation, peatlands drained for afforestation may continue to be a source of C in spite of forest biomass growth. Thus drainage and afforestation of peatlands reduces CH_4 emissions and provides a larger CO_2 sink than the peatland plant community.

However, the net exchange of C by peatlands is an upward flux for some years, until the C fixed by the canopy exceeds the losses of C from the oxidizing peat with the lower water table. The primary interest in forest-atmosphere C fluxes is the contribution to the global atmosphere C budgets and the consequent radiative forcing of climate. As CH_4 on a per molecule basis in radiative equivalents has a global warming potential 30 times that of CO_2, it is not simply the mass fraction that should be used to identify priorities. The change in the net direction of CH_4 flux from emissions to deposition needs to be included in C budgets and the overall net radiative forcing calculated.

Future Carbon Sequestration Potential

Across Europe forests are now growing faster than previously and similarly in the tropics. We currently seem to be in a period of especially strong C uptake by forests, although it has been suggested that this will not continue indefinitely, and indeed there is concern that the rate will diminish significantly as the 21st century proceeds.

It has frequently been suggested that the present rates of CO_2 uptake and growth are the combined result of the recent global increases in atmospheric CO_2 concentration (*e.g.* Jarvis 1998), N deposition and temperature. Quantitative evaluation of the relative impacts of all three of these factors is a focus of current research.

CO_2-stimulated Carbon Sequestration

It is difficult to measure the impact of the slow progressive rise in atmospheric CO_2 concentration on the C sequestration capacity of stands of trees but some relevant indications can be obtained from recent experimental programmes to evaluate the impacts of elevated atmospheric CO_2 concentration on the growth, development and physiology of young trees.

Doubling atmospheric CO_2 concentration speeds up development, photosynthesis and growth by up to 60% in a wide range of species of young trees grown rooted in the ground in the field over periods of up to 5 years in both North America and Europe. Moderate lack of nutrients and of water reduce growth but have little adverse effect on the *relative* impact of the increase in CO_2 concentration. To all intents and purposes the impact of the increase in CO_2 concentration is that trees get bigger quicker but otherwise remain similar in most respects to trees of the same size in the current conditions. Meta-analysis of the results from many experiments shows that the key physiological parameters defining capacity for photosyn-thetic CO_2 uptake by trees, V_{cmax} and J_{max}, are downregulated on average by no more than 12% and stomatal conductance by 15% as a result of growth in elevated CO_2 concentration.

A key question, as yet unresolved, is to what extent will the responses of key parameters to increased CO_2 concentration change when a stand of young trees reaches canopy closure. This is not readily solvable other than by long-term, free air elevated CO_2 concentration exposure experiments (FACE), starting with seedlings or clonal propagules. So far, the few FACE experiments in progress in forests have been imposed on existing stands, which are unlikely to acclimate fully over, say, a 5-year period of exposure. Thus major uncertainties remain as to the likely long-term effects of the rise in atmospheric CO_2 concentration on C sequestration and the contribution of CO_2 fertilization to the global terrestrial C sink.

Temperature-stimulated Carbon return to the Atmosphere

Photosynthesis is particularly dependent on solar radiation, CO_2 concentration and N availability and thus is likely to be reduced by increase in cloudiness but increased by elevations of the global atmospheric CO_2 concentration and by N deposition. All respiratory processes are very sensitive to temperature, as is the multiplication of the populations of respiring organisms, particularly the fine roots and heterotrophs in the soil.

Thus soil respiration increases strongly in boreal and temperate forests in relation to the rise in soil temperature as the spring progresses and consequently is likely to be very sensitive to the increase in global temperature. Increase in soil temperature leads to enhanced mineralization of soil organic matter and release of N, which feed back to stimulate photosynthesis, tree growth and transfer of CO_2 from atmosphere to forest. The balance between these conflicting effects of rising temperature on the net exchange of CO_2 (enhanced respiration and enhanced photosynthesis) is only now being resolved experimentally in the short term, although models indicate a potentially complex time-course in the long term.

Carbon Sink Saturation

It has been proposed that on time-scales of decades to centuries saturation of C sinks will occur and that existing C sinks will become significant C sources. This effect, it is suggested, arises from a difference in phase and functional relationship between the responses to global climate change of the principal gain and loss terms in NEE.

The CO_2 stimulation of photosynthesis increases with increasing atmospheric CO_2 concentration (the CO_2-fertilization effect) but does so at a diminishing rate. On the other hand, heterotrophic respiration (decomposition of litter and soil C, resulting in emission of CO_2 to the atmosphere) increases somewhat later in time as detritus production builds up and increases exponentially with increasing temperature.

Because so much C is stored in forest soils, and because this C is potentially vulnerable to climate change, it is essential to take full account of the C stocks in soil. It has been postulated on the basis of models that changes in climate, particularly the increase in temperature, will lead to higher rates of heterotrophic respiration and oxidation of the C stock in the soils, it is presumed without acclimation. Thus, as both atmospheric CO_2 concentration and temperature rise, the rate of heterotrophic respiration may be expected to increase relative to the rate of photosynthesis, and the overall capacity of the terrestrial biosphere to take up C from the atmosphere to diminish. However there are a number of possible feedbacks not included in these model analyses. These include the effects of rising CO_2 and temperature on:

- litter quantity, which may increase as a result of any short-term CO_2-fertilization effect and may lead to enhanced nutrient availability and C storage in soils;
- litter quality (*i.e.* the C:N ratio), which may decrease and retard decomposition and nutrient release;
- allocation of C below ground, which may well increase and lead to increased production of roots, mycorrhizae and exudates;
- increased nutrient availability and temperature leading to higher photosynthetic rates and NPP; and
- longer photosynthetic and growing seasons.

These feedbacks raise major uncertainties, currently being addressed by soil-warming experiments and modelling. The results of recent modelling analyses that have taken these feedbacks into account suggest that the dominant *short-term* constraint on the CO_2-fertilization effect is an increase in litter quantity. On the other hand, *long-term* increases in C sequestration are possible as atmospheric CO_2 concentration increases, if the net N input also increases (either through fixation or deposition) or if the N : C composition of newly formed litter substantially decreases. If below-ground C allocation increases, as observed in many experiments, and leads to increased production of roots, mycorrhizae and exudates, with consequent increase in N fixation, a large long-term increase in NPP and C sequestration is likely.

The simulated *long-term* effect of rising temperature is a large increase in NPP, and hence litter production, as a result of enhanced N mineralization and consequent increased photosynthesis, whereas *short-term* effects of temperature on photosynthesis and respiration are relatively unimportant.

These simulations highlight the importance of litter production on the one hand and turnover processes in the soil on the other, the former being strongly constrained by N availability and air temperature, the latter by soil microbiological processes and soil temperature. One cannot escape the conclusion that more emphasis needs to be placed on understanding better the processes in the soil, and the link between soil and atmosphere through the medium of the trees.

Indications from soil-warming experiments are that after an early initial flush of CO_2, the CO_2 efflux is similar in the heated and unheated control plots; furthermore, the feedbacks are actually increasing the efficiency of the 'forest pump' through the release of N and may override the initial losses of C. The effect of climate change on the efficiency of the pump in transferring C from the atmosphere to the soil via enhanced nutrient availability, photosynthesis and detritus production needs to be evaluated in addition to the effect of climate change on the return of C from the soil to the atmosphere, in order to obtain a complete picture of the net change in C sequestration by forests in a warmer climate.

A further key question is whether the current annual rate of N deposition (= 20kgha–^{1}) adequately balances the current annual rise in atmospheric CO_2 concentration (= 1.5mmolmol^{-1}) to sustain the NEE over the next 50 years. As things are at the present, the possibility of C sinks becoming sources in the mid twenty-first century remains highly speculative.

Forests of the Carboniferous

The Carboniferous saw the extensive development of lowland forest types, as we should expect of a period when the continents were generally at a low level. Again, however, we must presume a certain amount of upland, even with mountains of possibly considerable height, for there are far too many fragments of trunk and branch wood, with no associated leaves or other structures with which they can be related, to be accounted for in any other way.

Furthermore, where else but from adjacent mountains and extensive uplands could the sediments producing the abundant sandstones and shales of the time have been derived? But it is to the lowlands, with their rich fossil record, that we turn our immediate attention.

It has usually been thought that the extensive Carboniferous lowlands were characterized by extremely heavy precipitation and there is no doubt that swamps were widespread. What sort of "swamps" were they? The answer is to be found in the trees that inhabited them.

To live and function actively every cell must have a relatively abundant supply of oxygen and also be able to get rid of any excess of carbon dioxide produced during certain physiological reactions. Therefore, when we examine the plants of constantly wet areas, we find that those portions where the aeration is poor develop a sort of internal tissue known as aerenchyma.

This specialized tissue consists of a series of ramifying ducts, channels, or connecting loose-celled structures whereby the living cells of the submerged portions can exchange air with the parts in normal atmosphere.

There is little point in arguing, as some have done, that "plants were different then." Protoplasm was protoplasm, just as it is today and there is absolutely no evidence that it had any different requirements or tolerances. The forms of the plants have, perhaps, changed somewhat (fundamentally very little), but the basic physiology seemingly has not. Therefore, we can examine the roots of these groups of Carboniferous plants and tell with considerable precision whether or not the soil in which they grew was saturated with water.

Of the roots of Carboniferous plants that have been examined in any detail, only the following are known to have possessed aerenchyma: the equisetopsids in general (the sphenophylls and arborescent calamites), the marattiaceous tree fern *Psaronius* and perhaps a few, but by no means all, of the arborescent lepidodendrids.

The great bulk of the lepidodendrids and all the other arborescent forms of the time, so far as is known, were devoid of it, indicating that they did not inhabit areas which were flooded and waterlogged for considerable periods. But how, then, can we explain the great deposits of coal in the Carboniferous, if they were not laid down in vast, reeking swamps? Here again the Coal Measure strata tell the story.

These abundant plant remains accumulated in slowly descending basins. Near the estuaries, which often extended well into the continental interiors, in actual depressions inland and along streams, the equisetopsids, *Psaronius* and a few of the many species of lepidodendrids, with their specialized root aerenchyma, were abundant. With an equable and completely frost-free climate extending at least to latitudes 60° N and S, there would be ample opportunity for rampant growth over a large portion of the land surface if soil moisture were reasonably available.

In the slowly descending basins characteristic of large areas of the land surface at various periods of the Carboniferous the soil moisture was certainly ample at all times of the year. However, this does not mean that the rainfall was necessarily excessive or the soil saturated with water, except locally. With the generally low relief of these basins would have been associated a complex pattern of saturated, poorly aerated soils, intermingled with soils that were moist but had sufficient aeration for the growth of plants without a specialized aerenchyma in their roots. So intricate was the pattern and so dense the vegetation that as the trees fell, to be preserved in the deposits, they fell in such a way as to make it seem at first glance as if they had all grown in exactly the same habitat.

The same intricate pattern in the distribution of different species is easily discerned in our modern swamp forests, where we find some species capable of persisting with their roots submerged, while others, growing so close as to have their branches touching, may be seen to occupy slight rises on which the soil, although constantly moist, is much better aerated. These latter species also are likely to be found wherever the soil moisture is abundant outside the swampy zone, sometimes even on uplands.

Although moisture was reasonably constant in the lowland soils, the humidity of the air was not excessive. Careful study reveals that the great majority of the Carboniferous plants had leaves which in some manner resisted water loss through transpiration.

This applies to the arborescent lepidodendrids, the pteridosperms and the cordaites — the last a group of lowland pinopsid trees then in great abundance, which sometimes formed. dense stands. Even the microphyllous equisetopsid trees of the time, growing in their soggy environments, indicate that the atmosphere was by no means saturated with water vapour.

For our purpose, however, it is perhaps not overly important what the conditions were under which these lowland forests grew, except insofar as they shed light upon the worldwide atmospheric and climatic conditions of the time and thus give clues as to what was probably happening on the even drier uplands, where the ancestors of our contemporary trees already were undergoing their primary development.

Actually, we have little first-hand evidence regarding the upland forests of the Carboniferous, apart from what is revealed by the relatively few fragments of trunk and branch wood that floated into the lowlands, there to be preserved in the deposits.

While these give us valid hints, it is from the groups that migrated into the more equable lowlands during the severe climate of the Permian that we can glean a more satisfactory picture of the upland forests of the Carboniferous. Just as the wood fragments indicate, these migrants consisted mostly of a series of pinopsid forest types, with forerunners of the true "conifers" dominant among them. Often made up of giant trees (specimens with trunks up to six feet in diameter are known), these pinopsid forests probably had an understory of pteridospermous trees, shrubs and vines (with perhaps a few herbaceous types) and there is every likelihood that ferns of diverse sorts were abundant.

The subhumid areas near the 30° parallels probably were not devoid of vegetation. The tough-leafed pteridosperms of the time were already giving rise to the cycads and cycadeoids, both admirably fitted to less humid — even xeric — conditions.

Therefore, it is probable that the less severe parts of these subhumid areas supported a variety of pteridosperms suited to such conditions. And a few scattered fossils indicate that ancestral elements of certain lines, leading to those of the modern pinopsids better suited to dry areas, already were present. Cordaitean forests dominated the moist lowlands; ancestral "conifers" were beginning to take over the uplands.

If any forests grew in the cool-temperate regions and on the slopes of any higher mountains of the Carboniferous period, their nature is largely a matter of conjecture. If, however, the requirements necessary in the present are a mirror of those of the past, there is every reason to believe that abundant forests were then present in the Arctic and Antarctic regions, unless they were limited in some manner by the lack of light during the long polar winter; certainly the temperatures were no bar to the development of dense forest stands.

If so large a number of pinopsids can today stand temperatures far below freezing, surely the large and variable group then extant could have included forms capable of persisting in the polar regions in the less rigorous temperate climates of the time.

Their apparent abundance on upland habitats at middle latitudes during the period is an almost certain indication that the pinopsids also included a wide series of types adapted to temperate climates. And since resistance to frost in plants is physiologically closely akin to resistance to drought, there is considerable reason to believe that the poleward temperate regions of the time also supported at least a scattered assortment of pteridospermous types.

Although there is little or no evidence that the vast lowland forests of the Carboniferous contributed anything to the development of modern arborescent types, they played a most important role in relation to the use and ultimate preservation of our present-day forests.

Before the widespread use of coal, wood was almost the only source of fuel. In viewing our dwindling forest resources, we may be thankful that the lowland forests of Carboniferous time, entombed in the strata of the Coal Measures, have relieved our modern forests of much of this great and potentially destructive burden.

The Great Change in the Permian

The Permian opened with a climate very similar to that of the latter parts of the Carboniferous and with the same general distribution of forest types. This pattern was soon disorganized by what may have been the greatest period of glaciation that the world has ever experienced.

By comparison, it dwarfs any of the glacial cycles of the Pleistocene, which terminated the Tertiary. In the northern hemisphere evidences of Permian glaciation are known from Alaska, Massachusetts, England, France, Germany, the Ural Mountains and India. In the southern hemisphere all of what is now Australia was covered and extensive glacial deposits of the same age are reported from Madagascar, various parts of southern and southwestern Africa and South America, notably Brazil, Argentina and Bolivia. Nor is there any hint that these deposits were derived from local mountain glaciers; they were of a continental type.

The evidence of widespread glaciation during the Permian in the southern hemisphere, coupled with that of an amazing glaciation in what is now tropical India in the northern, did as much as anything to spark the far-from-settled controversy concerning the possibility of continental "drift." If, as postulated by the proponents of the theory, these glacial deposits could have been laid down in polar regions and then later transported into what ordinarily have been thought of as the temperate and tropical regions by a lateral shifting of the continental masses, there would be no need to disturb the "normal" climatic belts of the world.

The possibility of at least a certain amount of continental displacement need not be ruled out. But, the story told by the forest types of the time indicates that a vast world-wide change in climates actually did take place: those groups which, since the Devonian, had been limited chiefly to frost-free regions and apparently fixed in their evolutionary potentialities were essentially wiped off the face of the earth.

The climatic situation during the height of the Permian glaciation. The completely frost-free zone may have dwindled to a narrow band in the equatorial regions, thus cutting down a multitude of tropical genera and species.

The continental uplifts would have drained the great Carboniferous swamplands, thereby eliminating other arborescent types with which the world had been so long familiar and interior plateau deserts with severe climates would have been extensive. It was a time of crisis for all the old forms of life — animal and plant and many of them failed to meet the challenge.

Since there is little point in dwelling on what disappeared, let us turn to that which survived. Actually, the frost-free areas may have been rather more extensive than some students have supposed, for a few of the Carboniferous lepidodendrids, calamites, cordaites and pteridosperms passed through the perilous times of the Permian and lingered on into the Triassic. Furthermore, the fossils tell us that the land unoccupied by glaciers was never barren of vegetation except in the worst of the and areas and in the polar regions; there were vascular plants of one sort or another right up to the ice fronts. And the rapid development of tropical lowland forests in the Triassic also points to the presence, in the Permian, of considerable areas of at least warm-temperate climate.

We leave this most interesting and critical period in the world's vegetational history with only one last glance. Various modern types of fern had become more evident. The cycads and cycadeoids were now common and venturing into the drier habitats then widespread. The ginkgos, the true "conifers," and perhaps even the earliest of the taxads were abundant, not only in the lowlands, but also on those uplands where the climates were not too severe.

The pteridosperms, although on the wane, still displayed a wide variety of forms. As previously noted, among these were groups which, for want of a better term, might be collectively called the "aberrant pteridosperms" — plants often with entire margined leaves and a reticulate venation. The male reproductive cells, or sperm, still were motile and for this reason they have traditionally been classed with the pteridosperms, although recognized as distinct orders and families.

Are we here getting our first glimpse of the real ancestors of the magnoliopsids? If these considerably varied "aberrant pteridosperms" actually were an independent complex of forms out of which the magnoliopsids were derived, then the Permian marks the really important point of change in the floras of the world, a closing out of the old forms and a bringing into prominence of the basic members of all the important modern arborescent groups.

Nutrient Cycling In Forests

A nutrient cycle (or ecological recycling) is the movement and exchange of organic and inorganic matter back into the production of living matter. The process is regulated by food web pathways that decompose matter into mineral nutrients. Nutrient cycles occur within ecosystems. Ecosystems are interconnected systems where matter and energy flows and is exchanged as organisms feed, digest, and migrate about. Minerals and nutrients accumulate in varied densities and uneven configurations across the planet. Ecosystems recycle locally, converting mineral nutrients into the production of biomass, and on a larger scale they participate in a global system of inputs and outputs where matter is exchanged and transported through a larger system of biogeochemical cycles.

Soil Chemistry and Nutrient Vycling

By and large, the study of soil chemistry has concentrated on agricultural soils. The immediate practical questions to be answered have centred on the availability of plant nutrients, the need to develop chemical tests to define those soils suitable for particular crops and the prescription of the quantity of fertilizer that must be added to increase the productivity of the crop. With the exclusion of N (the availability of which is driven by biological processes in all soils), we might summarize the essential difference between the availability of plant nutrients in agricultural and forest soils:

- The availability of nutrients in agricultural soils is an immediate concern that mostly addresses inorganic equilibria over weeks or months;
- The availability of nutrients in forest soils is a long-term concern that should address biological processes and inorganic equilibria over years and centuries.

The growth of forests is a long-term process in which nutrients are cycled from plant to soil in litterfall and root turnover. Nutrients are withdrawn from tissues as they age and are translocated to actively growing tissues.

Timber harvesting takes away the aged tissues with relatively low nutrient concentrations, leaving the tissues with relatively high nutrient concentrations on site. In contrast, all of the nutrient supply for an annual crop is taken up from the soil within the few months the crop takes to reach maturity.

There is little cycling of nutrients from plant to soil. At maturity, the most nutrient-rich parts of the crop (seeds, leaves, storage organs) are harvested, leaving only the tissues with relatively low nutrient concentrations behind.

Thus, from the pioneering work of Ebermayer (1876) more than a century ago, much of the work on forest soils has been directed towards nutrient cycling, in particular:

- Quantifying the cycle of nutrients (uptake and return between plant and soil and retranslocation within the plant);
- Quantifying inputs and outputs, including nutrient removals in harvested timber.

While all of this is obvious, the science required is not obvious. Twenty years ago Stone (1979) concluded that despite a wealth of data on nutrient cycling processes for many forests, we do not have a sound experimental basis on which to assess the sustained productivity of forest ecosystems. Ten years later, Landsberg *et al.* (1991) restated that conclusion: 'Conventional chemical analyses to determine the nutritional status of the soil yield information that may have little relevance in calculating the capacity of the soil to provide nutrients to trees, and the complexity of uptake processes through complex, dynamic and poorly defined root systems is tremendous.' Rather than attempting to review ongoing research on the experimental and analytical basis for assessing sustainability of nutrient supply, we give only a factual account of what is known of the chemistry and cycling of the major nutrients. We do this in general terms, using data for a temperate forest of above-average productivity and we include a summary of the form, function, concentration and cycling of the major nutrients N, P, Ca, Mg and K.

Biomass

The above-ground mass of the forest (trees, under-storey and shrub layer) may reach 500t ha^{-1}. Litterfall *(A)* is 8tha^{-1} $year^{-1}$ but may reach 10tha^{-1} $year^{-1}$. If the litter layers (*Q*) weigh 30tha^{-1}, the decomposition rate of litter $k = A/(A + Q) = 0.21$ $year^{-1}$, and the half-life $t_{0\ 5} = 0.693/ft = 3.3$ years. This and subsequent analyses are based only on the above-ground stand. Data for root mass and root turnover are both more limited and variable, and have been intensively summarized. We have previously estimated a rate of root turnover for our temperate forest of above-average productivity of 2-3tha^{-1} $year^{-1}$.

Nitrogen

The concentration of N in leaves is greater than the concentration of any other nutrient, but much of this N is withdrawn from leaves before they die and fall as litter. Internal redistribution of N accounts for almost 30% of the total annual above-ground cycle. The half-life of N during litter decomposition is 3.5 years, a little longer than that for dry weight, so that N is immobilized during decomposition of the litter. The input of N in rainfall of 5 kg ha^{-1} $year^{-1}$ is more or less balanced by the output of N in streamwater. More than 95% or so of N in the soil is in organic form; organic N is mineralized to NH_4^+-N and then under certain conditions to NO_3^--N, a form freely available for uptake by plants. NH_4^+-N is held on negatively charged exchange sites of soils, but NO_3^--N is mobile within the soil and is the form in which N moves with drainage waters into streams.

In soils of natural forests of the world, the generally high C/N ratio of litter and soil organic matter precludes mineralization proceeding to NO_3^--N, so that the dominant form of N is NH_4^+-N. The cycling of N is therefore conservative, and N may be immobilized in litter of older forests (particularly in cooler-temperate areas) to the extent that the trees become N-deficient. Disturbance (treefall and the creation of gaps, fire, harvesting for timber) provides the conditions that result in a decrease in C/N ratio and the production of NO_3^--N, in which form N may be leached from the soil into streamwater. Rapid regeneration of the forest, including rapid uptake of NO_3^--N, following disturbance is therefore the key to the conservation of N.

Trace Elements (micronutrients)

The trace elements Fe, Mn, Cu, Zn, Mo and B have all played spectacular parts in agriculture and in tree plantations where plants have been introduced into new land. Applications of trace amounts (*e.g.* as little as 0.5kgha^{-1} of Mo) have corrected deficiencies that would otherwise have made agriculture or plantation forestry unproductive. We know of no such experience in natural forests, although there is little doubt that the distribution of many species may be determined by their ability to increase or restrict the supply of one or more of the trace elements. For example, concentrations of Mn in leaves and bark differ consistently and significantly between groups of the genus *Eucalyptus* but whether these differences are of significance in determining the distribution of species is unknown.

Phosphorus

Phosphorus occurs in highly mobile forms within plants and, like N, is withdrawn from plant parts during senescence; internal redistribution of P accounts for 50% of the total annual above-ground cycle. The half-life of P during decomposition is 4.1 years, *i.e.* P is immobilized throughout decomposition to a greater degree than N. A negligible amount of P comes in with the rain. P as $H_2PO_4^-$ is adsorbed by clay colloids and is then highly immobile in the soils so that losses of P to streamwater are negligible. However, in forest soils more than 50% of the total P in surface soils may be in organic form. Mineralization of organic P is therefore fundamental to ecosystem function, and P availability is 'determined by competition between biological and geochemical sinks for phosphate anions'.

Whereas methods for estimating rates of N mineralization have been developed and are in general application, there are no methods for routinely estimating the rate of P mineralization. The difficulty is that whenever $H_2PO_4^-$ comes into solution, there is a surface to adsorb it. This competitiveness between biological and chemical sinks for P remains a priority area of research in quantifying the long-term sus-tainability of P supply for forests.

Calcium, Magnesium and Potassium

Ca, Mg and K, and the non-essential element Na, are generally grouped as the major exchangeable cations in soil. An exchangeable cation is held adsorbed on a negatively charged surface of a colloid (both inorganic and organic). In this exchangeable form, the cations are not soluble but are brought quickly into solution and made available for uptake by exchange of protons from plant roots.

Ca is highly immobile within the plant, where it is incorporated within cell walls. Thus 80% of the total annual above-ground cycle of Ca is in lit-terfall. In contrast, K is highly mobile and K^+ easily leached from tree crowns by rain; leaching accounts for about 30% of the total annual above-ground cycle of K.

Roots and the Rhizosphere

Roots supply plants with water and nutrients and provide anchorage in soil. Most studies of roots in forests have been directed to the surface horizons, where fine roots (< 2 mm in diameter) are concentrated and where water and the bulk of nutrients are sourced. There are few studies of roots below 1 m, with only nine reported in the review of Jackson *et al.* (1996). In general our knowledge of root characteristics, including biomass, distribution and fine-root area, is poorly developed compared with the above-ground components of forests. Much of the recent interest in root distribution and turnover in forests has arisen primarily from concern for increased C flux from soils due to climate change.

Distribution, Mass and Turnover of Roots

A global analysis of root distribution in terrestrial biomes showed that boreal forests have the shallowest rooting profiles (80-90% of roots in the upper 30 cm of forest soils) and that temperate coniferous forests have the deepest rooting profiles (52% of roots in the upper 30 cm). When considered together, all temperate and tropical trees have 26% of roots in the top 10 cm and 60% in the top 30 cm. Root biomass of forests has been estimated at 2-5 kg m^{-2}, with the highest values in tropical evergreen forests. However, discrepancies between model estimates and measured characteristics of fine-root length and surface area point to the need for more data on root biomass and distribution, and for more careful measurement of fine root biomass in particular.

These analyses of root distributions and characteristics (biomass, surface area and nutrient content) among forests will enable improved modelling of water and nutrient uptake on a global scale.

They also provide the basis for improving global models of the impact of climate change on vegetation distribution and C sequestration in soil. When linked with models that predict the change in vegetation distribution and type, they will allow better prediction of the consequences of global environmental change.

The need for this type of analysis seems all the more urgent when it is considered that roots can account for more than 50% of net primary productivity (NPP) in forests and that the global pool of C in fine roots is 5% of the size of the atmospheric C pool. Because about half the global pool of C in fine roots is in forests, further study of factors influencing fine-root turnover is critical to the understanding of CO_2 flux between terrestrial and atmospheric C pools.

While roots are concentrated in the upper metre of soil in most forests, deep roots become important in forests subject to periodic drought because they provide access to water held deep in the soil. Carbon from these deep roots may contribute over decades to net C release from soil following conversion of forest to pasture, as shown by Nepstad *et al.* (1994) in an Amazonian rain forest. These sources of deep soil C have not been generally considered in estimates of C flux from forest soils, especially those of tropical regions where forests continue to be cleared for agriculture.

The turnover of fine roots has been investigated in relation to forest disturbance, including clearcutting, in the Northern Hardwood ecosystem. Biomass of fine roots (<2mm) in mature hardwood forest was 471 gm–^{2}, with average lifespan ranging from 8 to 10 months. Biomass of fine roots recovered to 71% of that in the mature forest within 3 years after clearcutting.

Rhizosphere

The rhizosphere is the area of soil immediately surrounding the root where biotic and abiotic processes interact to create an environment distinct from bulk soil further from the root. Compared with bulk soil, rhizosphere soil is usually more intensively weathered, has a lower pH and has greater concentrations of cations and P. Exudates released from plant roots and microorganisms benefit plant growth by increasing nutrient acquisition and metal detoxification, by alleviation of anaerobic stress in roots and by their action in mineral weathering.

There is no doubt that some forest species can modify the rhizosphere to access previously unavailable reserves of soil P by releasing organic acids that solubilize inorganic P. For example, Grierson (1992) demonstrated that in P-deficient soils, *Banksia integrifolia* induced the development of short-branched, tertiary, lateral roots (proteoid or cluster roots) that released citric and malic acids (the main organic acids induced under P deficiency).

Organic acids probably increase the availability of P in soils through a combination of decreased adsorption of P, increased solubiliza-tion of P compounds and chelation of metals such as Fe. Other rhizosphere processes that impact on the availability of soil P reserves include increase in root hair length and density, release of C that enhances mycorrhizal exploitation of bulk soil, and the release of phosphatases that solubilize organically bound P.

The role and function of mycorrhizas in forest soils has been reviewed by Brundrett (1991) and Vogt *et al.* (1991) and there is no doubt that mycorrhizal associations benefit trees by enhancing nutrient acquisition, especially in relation to P. The evidence for N and P transfer to forest trees via mycorrhizal associations has been demonstrated in tracer studies run over relatively short time-courses. However, it is only recently that new techniques have allowed longer-term studies; these have confirmed the benefits of mycorrhizas to nutrient acquisition and have also demonstrated increased plant growth as a result. Further work of this kind is required to establish the relative contribution of mycorrhizal uptake mechanisms to nutrient acquisition in mature forests over the long term.

The activity of microorganisms in the rhizos-phere is driven by the release of low-molecular-weight organic compounds by the root and there is some evidence to suggest that the growth of certain microorganisms is favoured while that of others is suppressed. For example, Gonzalez *et al.*(1995) reported that while microbial numbers were greater in the rhizosphere of *Alnus,* roots differentially favoured colonization by proteolytic and ammonifying organisms and inhibited nitrifying organisms. Norton and Firestone (1996) reported a greater than 50% increase in N turnover (mineralization and immobilization) in the rhizosphere relative to bulk soils in *Pinus ponderosa* microcosms. Further evidence of the complex interactions in the rhizosphere come from reports of ectomycorrhizal fungi stimulating bacterial growth associated with the release of citric acid by the fungus and, conversely, of inhi bition of bacterial growth.

Recent work has aimed at characterizing the chemical and biotic nature of the rhizospere, especially as it is influenced by soil liming and nutrient addition. Most work has been focused in the vicinity of industrialized areas of the Northern Hemisphere, where there are concerns for the unimpeded functioning of beneficial root-rhizosphere reactions under both controlled (fertilizer inputs) and uncontrolled (atmospheric deposition) inputs of acidity, N and S. Reported effects of ammonium sulphate application on rhizosphere chemistry vary and many of the changes, such as increased acidity and Al, are found in bulk soils as well. Other effects include depletion of P and K from the rhizosphere of Norway spruce in southern Sweden induced by ammonium deposition and the subsequent stimulation of P and K uptake.

The benefits of rhizosphere organisms, including mycorrhizal fungi, in maintaining the growth of tree seedlings in nurseries has long been known. This concept has been extended recently to the development of soil inoculation techniques to improve the growth of seedlings in the reforestation of degraded or clearcut sites , where inoculation with forest soil has increased the number of ectomycorrhizas. Differential effects on survival and growth of seedlings, where planting holes were inoculated with soil transferred from forestry or plantation sites, have been reported. Increases in seedling growth have been associated with increased nutrient mineralization brought about by soil animals stimulating microbial turnover in transferred soil, while increases in survival probably resulted from the introduction of beneficial rhizosphere organisms.

Another dimension in the use of mycorrhizas to benefit tree growth has been the recognition by Garbaye (1994) of bacteria that selectively promote mycorrhizal development (the so-called 'mycorrhization helper' bacteria, MHB). In anticipation of the applications discussed by Chanway (1997) the process of selecting and using MHB cultures for controlled 'mycorrhization' has been patented by Garbaye and Duponnis (1991).

Production Forestry and Carbon Sequestration Rationale for Carbon Sequestration

Climate change (generally alluding to global warming) refers to the variation in the climate of the Earth due to various factors such as variation in solar radiation, the Earth's orbit and greenhouse gas concentrations. Of these factors, increasing concentrations of greenhouse gases (predominantly carbon dioxide, CO_2) is attributed to anthropogenic activities. The carbon reservoir (carbon dioxide sink or CO_2 sink) is increasing in size, and the global carbon stock is projected to increase by a net 290 Gt (Gt = thousand million tonnes) due to CO_2 increases, climate change and vegetation redistribution.

Carbon sequestration describes the processes that remove the carbon from the biosphere. The IPCC concluded that the 'cumulative amount of carbon that could potentially be conserved and sequestered over the period 1995-2050 by slowing deforestation (138 million ha) and promoting natural forest regeneration in the tropics (217 million ha), combined with the implementation of a global forestation programme (345 million ha of plantations and agroforests) would be about 60-87 GtC-, equivalent to 12-15 percent of the projected cumulative fossil fuel and deforestation emissions over the same period'. Many other ways have been suggested to mitigate the increase of carbon in the atmosphere.

Land use, Land-use Change and Forestry (LULUCF) and agriculture sequester carbon in forests and agricultural lands. The other option is to decrease emissions of CO_2 and other Greenhouse Gases (GHGs) by substituting fossil fuels with biomass substitutes. While this would be a better way to control CO_2 emissions, forestation is a tool for bringing about a net reduction in atmospheric carbon levels.

Current Status of Carbon Sequestered in Indian Forests

The forestry sector has the potential to sequester carbon in addition to sustain its carbon. The total carbon pool of forests ~~are~~ is assessed from estimates of biomass from growing stock of forest stratum. IPCC IPCC (1995). Guidelines for greenhouse gas inventory workbook, Volume 2, Module 5 - Land Use Change and Forestry, Report prepared by UNEP, OECD, IEA and IPCC, 5.1-5.45. Prescribes a default value of 0.95 as the conversion and expansion factor to convert volume (in cubic metres) to biomass (expressed as tonnes of dry matter). IPCC also prescribes a factor 0.45 for converting the biomass into carbon content. Other researchers have used similar values, *e.g.*, Chaturvedi assumed the factor to be 0.48 and Bhadwal and Singh assumed the factor to be 0.5.

The major grouping of Indian forests (based on climate) are:

1. tropical
2. montane subtropical
3. montane temperate
4. alpine.

These are subdivided into 16 sub-types based on Champion and Seth classification. Of these sub-types, majority of the area is under tropical moist deciduous and tropical dry deciduous forests. The total forest cover of India has been around 64 Mha since the 1980s. Dhadwal and Nayak (As cited in Lal and Singh) estimated the total carbon pool of Indian forests to be 1994 million tonnes (Mt) for the year 1985. Ravindranath and Somashekhar estimated the total carbon pool for Indian forests to be 4179 Mt for the year 1986.

In a more recent research study, Lal and Singh estimated the total carbon pool for Indian forests for the year 1995 to be 2027 Mt. Dense natural forests are better stores of carbon than open natural forests. A study of natural forests of Madhya Pradesh by Pande estimated the total carbon pool in standing crop as 363 Mt for dense forests and 80 Mt for open forests. Open forests, therefore, have a potential to sequester more carbon.

It is also to be noted that the annual productivity of Indian forests have been reported to increase from 0.7 m^3 /ha in 1985 to 1.37 m^3 /ha in 1995.

Estimates of Future Carbon Sequestration

Prasad *et al.* have modeled the land use changes and forestry data of India from 1997 to 1999 and suggested that Indian forests would be a potential sink for 0.94 Gt of carbon over a period of time, with an increase in dense forest area of about 76 Mha and decrease of about 3 Mha and 5 Mha in open and scrub forests, if similar land use changes that occurred during 1997-1999 would continue. The study bases its observations on final outcomes on the land use patterns (and changes therein) observed between 1997 and 1999. The authors use stochastic modeling techniques to determine relationships between land usage and magnitude of carbon stored in forests and soils. They extend the model to map net carbon flux between the land and the atmosphere. An entire series of studies on methods for examining the future consequences of changes in land usage and carbon flux patterns are also referred to and evaluated as sources of suggested carbon mitigation strategy options.

A study by Bhadwal and Singh uses the Land Use and Carbon Sequestration (LUCS) model with software developed at the World Resources Institute, Washington, DC, USA. Indian agriculture and forest statistics were systematically integrated with geographic and demographic data. Three distinct scenarios – business as usual, conservation and plantation forestry were considered. It was estimated that under plantation forestry, 7 billion tonnes of carbon will be sequestered during 2000-2050.

In an earlier study by Ravindranath *et al.*, a sustainable forestry scenario and a commercial forestry scenario were developed with the aim of meeting the biomass demands through plantation forestry and conserving the forests.

Comparison of the commercial forestry scenario over the baseline scenario (in the period 2000-2012) estimated that an additional carbon stock of 78×10^6 Mg C would be sequestered apart from meeting all the incremental biomass demands (estimated for 2000-2015).

Economics of Production Forestry

In the study mentioned above, Ravindranath *et al.* considered the cost-effectiveness of four potential forestry activities to meet the biomass demand of the country. These activities were:

(1) short-rotation forestry (afforestation to meet the fuelwood and industrial wood requirements),
(2) long-rotation forestry (afforestation to meet the structural (sawnwood) requirements),
(3) forest regeneration (reforestation to regenerate and reclaim the degraded forest lands through protection and promotion of forest succession), and
(4) forest protection (to conserve the biomass of natural forest by halting deforestation).

The mitigation potential on a per-hectare basis for the period 2000-2030 was lowest for short-rotation forestry (at 25 Mg C/ha) and highest for forest protection (at 176 Mg C/ha). The life cycle cost of mitigation was lowest for forest protection (USD 0.39 /Mg C) followed by forest regeneration (USD 0.99 /Mg C), long-rotation forestry (USD 11.69 /Mg C), and short-rotation forestry (USD 18.8 /Mg C).

Status of Production Forestry in India

The area under plantation in India has been growing over the past decades. The rate of afforestation in India was 1.67 Mha/yr during the 1980s and 2 Mha/yr in the 1990s. Agroforestry also has the potential to store carbon and remove CO_2 through enhanced growth of trees and shrubs and has been demonstrated as a promising mechanism of carbon sequestration in India. Carbon sequestration in Indian agroforests varies from 20 tC /ha/yr in Uttar Pradesh to a carbon pool of 23-47 tC/ha/yr in tree-bearing arid agroecosystems of Rajasthan. The average sequestration potential of agroforestry has been estimated to be 25 tC/ha/yr over 96 million ha of land in the country.

With the aim to sequester carbon rapidly, it is best to grow fast-growing species with shorter rotation periods. In India, the choice of species for plantations is dictated by market demand for eucalyptus and poplar poles. Other fast growing species are *Albizia*, *Cassia siamea*, *Leucaena leucocephala, Casuarina equistifolia*, and *Dalbergia sissoo*.

Afforestation in India is generally carried out under the social forestry programme. This includes growing trees on community commons, public or government owned land and aveanue plantations. Farm forestry that includes raising plantations of eucalyptus, casuarina, teak and other commercial species on farmland, or on bunds is considered agro-forestry. Eucalyptus species and *Acacia auriculiformis* are grown under the reforestation programmes that included the following two categories of plantations:

1. plantations on village commons and degraded forest land that were carried out to meet the biomass need, such as for fuel and small timber.
2. farm forestry on farmers' own land to meet industrial demand

In India, plantations can also be categorized as:

1. short rotation plantation forestry (species such as *Eucalyptus*, *Acacia auriculiformis* and *Casuarina equisetifolia*), and
2. hardwood plantations (species such as *Tectona* and *Pinus*).

Broadly speaking, the major plantation species are poplar and eucalyptus in north India, subabool in central India, and eucalyptus, *Acacia auriculoformis* and rubberwood in south India. The productivity of forest plantations under the national afforestation programme is significantly higher compared to natural forests. A national level study of forest plantations in India estimated the mean anuual productivity (of woody biomass) to be 3.2 tonnes per hectare per year (t/ha/yr).

There are a few location-specific studies on productivity of forest plantations and the range of productivities recorded vary from 1.2 to 8.2 t/ha/yr, with the national mean productivity of about 3.1 t/ha/yr. This variation is not surprising since the case studies were location-specific and ranged from farm forestry in semi-arid regions to plantations on degraded forest lands in heavy rainfall zone. It productivities could range from 2-5 t/ha/yr in semi-arid regions and 5-10 t/ha/yr in sub-humid regions. One should also note that the productivity of any forest plantation would depend on the age of the trees.

TEMPERATURES AND CONDITIONS OF SANDY SOIL

Temperature

Soil temperatures are affected by a forest cover in the same way as air temperatures but to a much greater degree. Mean daily maximum temperatures at the soil surface may be reduced in summer by as much as 50° F. and absolute maxima even more. The influence of the forest varies inversely as the distance from the surface, but under some conditions may be recognizable to a depth of 30 feet.

Minimum temperatures show a similar but less pronounced increase. This influence is particularly noticeable in the winter, when soils usually freeze later and less deeply in the forest than in the open. Sometimes the soil in the forest remains unfrozen while that in adjacent open fields is frozen to a considerable depth. On the other hand, in situations where there is a heavy blanket of snow in the open and a very light blanket under a dense forest canopy, this relationship may be reversed. These influences are due partly to the reduction of insolation and radiation by the overhead forest cover and partly to the insulating effect of the litter and humus of the forest floor. These reduce maximum temperatures more than they increase minimum temperatures and hence the net effect is a cooling one.

In both cases the effect is beneficial. During the summer young plants are protected from high temperatures which in exposed situations in the open may be so extreme as to kill the living tissue; while during the winter reduction in the depth and period of freezing diminishes surface runoff by permitting the penetration of more water into forest soils.

Moisture Content

Forests tend to decrease the amount of moisture in the soil by intercepting precipitation, by retaining water in the forest floor and by transpiration. They tend to increase the amount by reducing surface runoff, by increasing the permeability of the soil and by decreasing evaporation from it.

Much precipitation in a forest never reaches the ground because it is first intercepted by and then evaporated from, the leaves, twigs, branches and trunks. This loss may run as high as one tenth of an inch per shower. The amount of water that actually reaches the ground may vary from none in light showers to the great bulk of the total precipitation in heavy rains. Interception naturally increases with the density of the stand and of the foliage and is much greater with deciduous trees in summer than in winter.

When the precipitation reaches the ground it is first absorbed by the forest floor, which has a field moisture capacity (that is, an ability to retain water against the pull of gravity) of from one to five times its own dry weight and when saturated contains much larger quantities. Its absorptive and storage ability is much greater than that of mineral soil, the field capacity of which usually ranges from 10 to 50 per cent of its dry weight. The thicker and the more decomposed the litter and humus of the forest floor, the more water it will hold.

The amount of water that actually gets through to the mineral soil may vary from zero to 100 per cent, depending on the amount of precipitation and on the condition of the forest floor. With light showers in times of drought all of the precipitation may be retained by the forest floor; while with heavy showers and in wet periods little or none may be so retained.

On land without vegetative cover and particularly with heavy soils, the surface layer tends to become hard and impermeable, with marked reduction in the amount of precipitation that sinks into the main body of the soil. In the forest, on the other hand, the litter slows down surface runoff and thus gives more time for the water to be absorbed by the humus, which acts as a huge sponge.

More important still, the humus prevents muddying of the underlying mineral soil, which remains porous and thus absorbs and retains much more water than does exposed soil in the open.

Enormous quantities of water are removed from the soil by transpiration. A well-stocked forest of mature trees uses more water than does any other form of vegetation. It may require from 100 to 1,400 pounds of water for every pound of dry matter produced.

Transpiration is far greater in summer than in winter and during the growing season is somewhat greater with broadleaf deciduous trees than with conifers. Its tendency is obviously to reduce the moisture content of soil in the forest as compared to that in the open. On the other hand, the reduction of evaporation in the forest, which was mentioned in the discussion of atmospheric conditions, tends to maintain the moisture content at a higher level than in the open.

The net effect of these conflicting influences varies widely with the character of the topography, the soil, the climate and the forest. So far as any generalization is possible, it may be concluded that soils under a forest cover tend to be somewhat drier than similar exposed soils in the open on fairly level ground and somewhat moister on steep slopes where more water is lost by surface runoff in the open than by interception and transpiration in the forest.

Erosion and Stream Flow

Of all the influences of the forest on its environment, the most obvious and the least controversial is the reduction of surface runoff of water. This prevents or substantially checks soil erosion and modifies stream flow.

Composition and Structure

Forests have a marked influence on the composition and structure of the underlying soil. Every year they add to the forest floor large quantities of leaves, twigs and branches, which under normal conditions are constantly decomposing to form humus. Part of this humus gradually mixes with the mineral soil beneath and also supplies it with soluble compounds carried downward by percolating water.

The amount of organic material in the forest floor varies widely with differences in climate, soil and vegetation. Recorded data run from less than 2 tons per acre in old-growth longleaf pine in Florida to nearly 120 tons in a forest of birch, sugar maple and spruce in New Hampshire. In general, high temperatures result in its rapid decomposition and disappearance. Ordinarily there is a gradual transition from undecomposed litter through partially decomposed duff and humus to a mixture of organic and inorganic material and eventually to mineral soil.

An exception occurs in coniferous forests in the north where slow decomposition of the needles results in acid "raw humus." This raw humus does not mix well with the underlying mineral soil and in extreme cases may be peeled off like a blanket.

The usual effect of a forest cover is to improve both the chemical and the physical characteristics of the soil. Nitrogen, calcium, phosphorus, potassium and other elements are made available in larger quantities. Even more important is the action of organic materials in making the soil more friable and crumbly. They tend to make light soils, such as sands, heavier; and heavy soils, such as clays, lighter. These effects are strengthened by the constant growth and death of the tree roots. While growing they keep pushing into and loosening up new areas of soil; when dead they add organic material to the soil and leave channels through which water may percolate more readily. Porosity of frozen soil is also increased in the forest, where it is more permeable than in the open.

Accelerated Erosion

Soil is constantly being moved by water and gravity from higher to lower elevations. Under normal conditions, with an undisturbed vegetative cover, this is a slow process known as "geologic" erosion. Over the centuries it has built up fertile alluvial soils in the valley bottoms. With the removal or disturbance of the vegetative cover, a striking change takes place. Larger and larger particles of soil and even huge boulders, can now be moved and the rate of removal is greatly increased.

The beneficial process of geologic erosion is replaced by the destructive process of "accelerated" erosion. Among the harmful results are deterioration or ruin of the lands where the erosion takes place and where the coarse detritus is deposited, siltation of reservoirs, impairment of the quality of water for municipal and industrial uses, destruction of fish habitats, clogging of river channels and increase in the volume of floods.

All kinds of vegetation serve to check erosion, but by far the most effective are well-stocked forests and well-sodded grasslands. Their influence is primarily due to their ability to reduce the amount and velocity of surface runoff.

The amount of material that water can carry at a given velocity is directly proportional to its volume. As the velocity increases, however, there is a much more than proportional increase in the cutting and carrying power of a stream. Thus, if the velocity of water is doubled, its cutting power is increased fourfold, its carrying power thirty-two-fold and the size of the material it can carry sixty-four-fold. These facts explain the occurrence of "mud-rock" flows, the transportation by small streams of boulders weighing many tons and up to a six-thousand-fold increase in the rate of erosion following destruction of the forest and the forest floor.

During ordinary storms much, if not all, of the precipitation is absorbed by the forest litter and humus, retained a short time and then passed on gradually to the mineral soil beneath. The humus layer is not comparable, as has sometimes been alleged, to a blotter on a slate roof, the efficacy of which in storing water ceases as soon as it becomes saturated. Except in "raw humus," there is no sharp line between the organic surface and the inorganic subsurface material in the forest.

On the contrary, there is a gradual transition from the top layer of newly fallen, undecomposed litter to partially and completely decomposed humus, to a decreasing mixture of organic with inorganic material and finally to mineral soil. Thus there is a constant and uninterrupted downward movement of water from the temporary storage provided by the upper layers of the forest floor into the larger and more permanent soil reservoir beneath. From there the water moves slowly as a subsurface flow to springs, streams and lakes, from which it is in time evaporated and again precipitated in the course of the hydrologic cycle.

Tree roots also help to keep the soil from being washed away in gullies and along the banks of streams and tend to check the occurrence of landslides. Erosion by wind is virtually nonexistent in the interior of a forest and is greatly reduced for some distance to its lee.

In many ways forests are influential in preventing the development of accelerated erosion — the greatest threat to the world's soil resources. Fortunately their influence is most effective where the danger is most acute, as on steep slopes, in clayey soils and with heavy rainfall. Wind erosion can also be prevented or greatly reduced by properly placed windbreaks in open country, as in the Great Plains of the United States.

Floods

What effect these influences of the forest have on floods is a hotly debated question. Certainly forests cannot prevent an abnormally high flow of streams when heavy precipitation falls on saturated or frozen soils. If the storage basin is already full, there is nothing for any additional supply of water to do but to flow off by way of the nearest stream. On the other hand, when the soil reservoir is well below the saturation point, there may be far less surface runoff from forested areas than from those not covered by vegetation, even in severe storms.

By and large there can be no doubt that forests tend to reduce flood flow in small watersheds in hilly country. One would expect them to have a similar influence on large rivers made up of the flow of many small streams. Here, however, the situation is complicated by many factors, the combined effect of which it is difficult to evaluate. For example, the way in which precipitation is distributed over a river basin as a whole and the rate of flow of the tributary streams largely determine whether their peak discharges reach the main river simultaneously or in successive waves.

Obviously the first combination will result in a much higher flood than the second, even with the same or perhaps even a smaller total runoff. Therefore, the clear-cut influence of the forest on the flow of streams from small watersheds at the headwaters of large rivers may not always be reflected in the flow of the main river.

Nevertheless, the general tendency will be for the forest to decrease both maximum and total runoff and to increase minimum runoff, throughout the river basin. The action of the forest in reducing erosion also has an important influence on flood flow and flood damage.

Small streams with steep slopes issuing from deforested or burned-over watersheds may carry several times as much solid material as water. One instance is on record in southern California where solids comprised 88 per cent of the total flow. As the stream flattens out, the larger and heavier materials are, of course, deposited; but even rivers with small gradients may carry heavy loads of sediment. This material adds significantly to the volume of the stream and increases its erosive power, particularly in time of flood. It may, in fact, be even more important than the volume of water as a source of damage.

Distribution of Runoff

For all practical purposes the total stream flow from any watershed consists of the precipitation less the sum of the losses from interception, transpiration and evaporation. To the extent that the forest increases interception and transpiration, it tends to reduce stream flow; and to the extent that it decreases evaporation, it tends to increase stream flow. The net effect depends on the relative weight of these influences. It will obviously not be the same under all environments or at all seasons of the year.

Measurements from forested and from deforested or partially forested areas in Switzerland, Colorado and North Carolina have shown a strong tendency toward smaller annual runoff from the forested areas. The difference was particularly marked in the spring, when melting snow and heavy rains caused the maximum seasonal runoff. The greater discharge of water from the nonforested areas also carried with it much more eroded material.

The total flow of a stream is made up of two main parts: the surface runoff and the subsurface or ground-water runoff. The forest has a profound effect in reducing surface runoff and thereby increasing the amount of water available for subsurface runoff. The evidence is conclusive that in most regions surface runoff is very small or negligible from areas of undisturbed vegetation, while it may amount to half of the precipitation where the vegetation has been destroyed or seriously disturbed. However, the outstanding influence of forests in this respect is greatly weakened when the forest floor is destroyed by repeated fires, even though the trees themselves are not killed.

Reduction of surface runoff results in more uniform flow of a stream throughout the year. Maximum flow is nearly always less and minimum flow usually more from forested than from comparable nonforested areas. The tendency is to avoid sharp peaks of flow in the spring and deep troughs in the summer. The total amount of flow may be less, but its greater uniformity is highly desirable.

Watershed Management

From the point of view of water supply, the ideal management of the forest, as a part of the broader field of watershed management, is that which will produce the maximum total runoff, well distributed throughout the year, with minimum erosion.

Such management may not always be identical with that which will produce the largest supply of wood. Serious conflicts will, however, be the exception rather than the rule. When they do occur, the method of management to be selected must be based on relative values, among which water for domestic and industrial use, irrigation, power, navigation and recreation may rank high. Striking evidence of the basic importance of conserving both soil and water resources is afforded by the greatly increased attention being paid by such agencies as the Soil Conservation Service, the Tennessee Valley Authority and the Forest Service in the United States to the promotion of more effective watershed management, in which forest planting and improved forest practices play a prominent part.

Wood, the Universal Raw Material

No less prominent than the influence of the forest on climate and runoff is its influence on industry. In fact, its tangible products affect our economic activities even more clearly and more directly than do the intangible services that it renders. They provide opportunity for the profitable employment of labour and capital in many important industries and furnish a wide variety of indispensable consumers' goods.

Wood is sometimes known as "the universal raw material" because it is so widely distributed and can be used for so many purposes. Although for some uses it has been largely displaced by other materials, there are few articles into the composition, manufacture, or transportation of which it does not enter. While perhaps no more "indispensable" in modern civilization than many other materials, such as iron, cement and glass, its range of use is certainly much broader than most of them.

For thousands of years, until the advent of coal, oil and gas, wood constituted man's chief source of fuel. Even today more than half of the total consumption of wood in the world goes for this purpose. Another third in the form of lumber and structural timbers is used for the construction of buildings of all kinds. Except in urban centers and where forests are scarce or lacking, most people continue to live in wooden houses.

Lumber and dimension stock are in turn remanufactured into innumerable articles such as toothpicks, toys, sporting goods, musical instruments, handles, boats, caskets and furniture. In the United States approximately an equal amount of lumber (about one seventh of the total cut) goes into boxes, crates and dunnage used in the shipment of commodities. Large additional amounts of wood are used for railroad ties, poles, piling, fence posts, mine timbers, excelsior, shingles, cooperage, veneer and plywood. Modern adhesives are greatly helping to expand the use of wood in the fields of ply and laminated construction and are making practicable the salvage of much low-grade material. Advances in engineering techniques now make possible the use of wood in members requiring high strength and long spans, where metal was formerly regarded as indispensable. Even more spectacular advances have taken place in the chemical utilization of wood. The demand for paper and paper products seems unlimited. Each year larger and larger quantities of wood go into the manufacture of newspaper, book and magazine paper, writing paper, wrapping paper, wallpaper, building paper, pasteboard cartons and paper board. From "dissolving pulp" come cellophane, rayon, artificial wool and many plastics. Today one could, if one chose, be clothed wholly in textiles that originated in the forest.

A lively imagination is required to visualize the many other chemical substances that can be derived from wood. Among these are wood gas, potash, charcoal, acetone, methyl alcohol, sugar, ethyl alcohol, yeast, salicylic acid and vanillin. Potentially the forest is the source of large quantities of motor fuel and food.

During World War II, when gasoline was scarce, many automobiles in Europe were operated entirely on wood gas, which is still used to a considerable extent in trucking. Another potential substitute for gasoline is ethyl (grain) alcohol produced by the fermentation of sugar, which in turn has been produced by the hydrolysis of wood cellulose. The alcohol can also be converted into lubricating oil and synthetic rubber.

Wood sugars have already been used extensively in Europe as feed for livestock and are a possible source of food for human beings. They can also be turned into yeast of high nutritive value by inoculating them with the appropriate organisms.

Edible carbohydrates and proteins are among the innumerable products obtainable from wood cellulose, which in its original state is a highly indigestible product. Lignin, the other main constituent of wood, is a comparatively unknown substance, of which little use is now made but which is likely in time to provide a wide variety of important chemical products.

Many substances other than wood and its derivatives come from trees and are ready for use with relatively little further processing. Among these are nuts, maple syrup, chewing gum, tannin, dyes, rubber and resin. The last yields turpentine, universally used as a solvent in paints and varnishes and in the production of synthetic camphor; rosin, used by violin players and baseball pitchers and in the manufacture of soap, varnish and paper; pine oil, used in the flotation process of separating metals from their ores and in the textile industry in the fixation of dyes; and pitch, for which chemists are just beginning to find profitable uses.

Many pages would be required merely to list the useful substances that come from trees. Wood, once regarded by some as obsolescent, is now used more widely and for more purposes than ever before. Its inherently desirable properties of high strength in relation to weight, its workability, elasticity, nonconductivity of heat and electricity and beauty can be enhanced and its undesirable properties of shrinking and swelling, inflammability and susceptibility to attack by insects and fungi can be mitigated by methods developed by modern technology. Wood as wood, properly treated and handled, is one of the most useful materials known to man; while cellulose, lignin and resin are a veritable treasure house of chemical products already in use and yet to be discovered. The significance of forest products is emphasized further by the facts that forests are both renewable and more efficient as producers of solid substance (ligno-cellulose) than is any other crop. Here is at least one case where we can have our cake and a rich cake at that and eat it too. Industrially the age of the forest is ahead of us as well as behind us.

Forest and Wood-Using Industries

This situation makes it possible for man, through intelligent handling of the forest resources, to raise his standard of living by the continued and increasing production of a host of valuable, often indispensable, raw materials and finished commodities. It also provides widespread opportunities for the profitable employment of land, labour and capital.

Tropical Sandy Soils

Sandy soils are characterized by less than 18% clay and more than 68% sand in the first 100 cm of the solum. In the World Reference Base (WRB) soil classification system, sandy soils may occur in the following Reference Soil Groups: Arenosols, Regosols, Leptosols and Fluvisols. These soils have developed in recently deposited sand materials such as alluvium or dunes. They are weakly developed and show poor horizonation. Soils characterized by a high proportion of sand in the first 100cm can also correspond to the upper part of highly developed soils formed in weathered quartz-rich material or rock, as evidence by the development of a highly depleted horizon. In the following discussion consideration will be given to a range of soils including sandy soils of the WRB and those with sandy horizons in the upper 100cm of the profile.

Sandy soils are often considered as soils with physical properties easy to define: weak structure or no structure, poor water retention properties, high permeability, highly sensitivity to compaction with many adverse consequences. However, analysis of the literature shows that their physical properties are far from simple. This is particularly true in the tropics where sandy soils are subjected to a cycle of wetting and drying that greatly affects the soil with small differences in composition leading to significant differences of physical properties.

Structure, Porosity and Bulk Density

Sandy soils are characterized by a lack of structure or that it is weakly development. Coquet (1995) measured the shrinkage properties of two soils in Senegal with different texture. On the sandy soil, results obtained in the field and in the laboratory (on cores originating from the same horizons), showed very small shrinkage: bulk volume variation was only 0.05%. When they dried, sandy soils develop very few thin cracks organised in a loose network. The meagre shrinkage properties of these soils are related to the low clay content and the high proportion of low activity clays of many tropical sandy soils.

A Large Range of Porosity

Sandy soils in the tropics show a large range of porosities and consequently bulk density (D_b). Porosity ranges from 33% (D_b = 1.78 g cm^{-3}) to 47% (D_b = 1.40gcm $^{-3}$) are commonly recorded. The porosity in sandy soils is usually smaller than in clayey and silty soils. Very small changes in porosities are generally observed in sandy soils of the tropics.

Lamotte *et al.* observed a porosity of 28% (D_b = 1.91 gcm $^{-3}$) between 35 and 45 cm depth in the Northern Cameroon's in very old cultivated fields. Lesturgez (2005) measured a porosity of 28% between 20 and 30 cm depth in a soil belonging to the Warin-Satuk series in Northern Thailand in cultivated soils. Deeper in the soil, similar small porosities were recorded by Burt *et al.* (2001) in sandy soils developed in a saprolithe derived from granitic rocks in Zimbabwe.

These small porosities were recorded in sandy soils with no gravel or stones thus indicating a close packing of elementary soil particles in soils that have been subjected to continuous cultivation.

However, under native vegetation with intense biological activity or after recent tillage operation (wheel tracks excluded), greater porosity of 60% (D_b = 1.10 gcm $^{-3}$) have also been recorded. Such a large porosity is related to the presence of numerous macropores that results from both faunal activity and root development.

Bruand *et al.* (2004) have also observed greater porosity in the subsoil of intensively cultivated soils, this being related to a loose assemblage of elementary soil particles unaffected by farming practices.

Significance of Sand and Silt Grain Size Distribution

Porosity varies with time after tillage operations thus making it difficult to attribute to soil composition alone. Osunbitan *et al.* (2005) showed a continuous decrease in the porosity of the 0-5 cm layer of a Nigerian loamy sand soil. Porosity ranged from 47.7% (D_b = 1.30 gcm $^{-3}$) to 60.4% (D_b = 1.05 gcm $^{-3}$) according to the tillage system and time after tillage. If we exclude the topsoil horizons from the dataset, the finesand: coarsesand ratio ranges from 0.5 to 6.1 for the data collected in the literature and the porosity tends to decrease when ratio increases (R^2 = 0.40, n=55). The fine sand particles occupying the voids resulting from the packing of the coarse particles would result in the porosity decreasing when the proportion of fine sand particles increases up to a value that would correspond to the total infilling of that void. With a greater proportion of fine sand, the porosity would start again to increase.

An increase in the silt-sand ratio wouldalso result in a decrease in the porosity as discussed by Agrawal (1991) for Indian loamy sand and sandy loam soils. These data recorded with soils samples are consistent with those obtained earlier with models and artificial mixtures in the laboratory. Fiès *et al.* (1972) studied the porosity of granular binary mixture and modelled porosity according to the proportion of coarse and fine fractions. They showed that for a mixture of 20-50 μm material with grains 2 mm in diameter, the porosity is minimum (P H"0.20) for 20-50 μm material content close to 25%. That proportion of fine material is consistent with the theory developed by Westman and Hugill (1930). They calculated an optimum ratio of 3.46 parts by mass of coarse sand to one part of very fine sand (*i.e.* 22.4% on mass basis of very fine sand) was required to obtain a mixture with the lowest porosity. Fiès and Stengel (1981) showed good concordance between porosity measured on small aggregates resulting from soil fragmentation and theoretical porosity computed with a model of binary mixtures. In particular, they showed that the porosity was at a minimum for a mixture of 2-20 μm and 200-2,000 μm when 2-20 μm content was close to 20%. These data indicate that the loose relationship between the porosity and the fine sand and coarse sand ratio would be valid for a limited range of fine sand and coarse sand ratio.

Role of the Clay Fraction Characteristics

The large range of porosity is related to the small cohesion forces between elementary particles thus enabling the formation of a large range of assemblages from very loose to very compact. This is specific to sandy soils because of the small amount of clay that can act as inter-grain cement. In the tropics, clay is often low activity clay (mainly kaolinite) and for similar clay content, tropical sandy soils show usually much smaller inter-grain cohesion than sandy soils in temperate and Mediterranean regions. In these deeply weathered soils, the clay-sized fraction may in part consist of quartz as observed by Hardy (1993) in soils that developed in sandy colluvial deposits in Northern Vietnam. Hardy (1993) showed that 10 to 40% of the <2 μm fraction was quartz in the soils studied. Bruand *et al.* (2004) studied sandy soils belonging to the Nam Phong series in Northeast Thailand and found that 25 to 35% of the <2 μm fraction was quartz.

The presence of quartz in the <2 μm fraction contributes to its low activity. In some sandy soils, however the presence of smectitic clays can lead to very different physical soil properties. In the semi-arid tropics, Lamotte *et al.* (1997) studied soil hardening in sandy soil with contrasting loose topsoil and underlying hard horizons. The horizons had similar particle size distributions and the hardness was closely related to a fabric with clay coatings on the sand grains and clay wall-shaped bridges linking the latter. This induced a strong continuity of the solid phase with only a minimum clay content of 6%.

Surface geology and distribution of sandy soils

Mesozoic sandstone dominates most of the basement geology in Cambodia and hence will have a dominating influence on the properties of upland soils. Recent and Pleistocene alluvial/colluvial and lacustrine sediments that now form the parent material for most of the lowland agricultural soils of Cambodia are substantially derived from the weathering and erosional products of the Mesozoic sandstone. However, low hills from felsic igneous intrusions particularly in South and Southeast Cambodia have also supplied siliceous sediments for the recent and older alluvial/colluvial terraces. In the Northeast of Cambodia, basaltic lava flows from the Pleistocene covered significant areas of older alluvial terraces.

The soils formed on weathered basalt and on the alluvial/colluvial sediments derived from basalt have very different properties to those of the siliceous parent materials that dominate most other soils. In the West of Cambodia, bordering Thailand substantial areas of siltstone limestone and marl occur, and this area is emerging as significant for upland crop production. Finally the sediments deposited by the Mekong River along its flood plain and in the basin of the Tonle Sap have resulted in a large part of Central Cambodia being dominated by recent alluvial/lacustrine sediments derived in part from the Mekong River basin and in part from the immediate basin of the Tonle Sap.

No specific mapping of sandy soils has been undertaken in Cambodia. A soil map (1:250,000) of most of Cambodia was recently completed based on the FAO World Soils Map (1988) as part of a soil resources map for the lower Mekong Basin. Parts of Cambodia fall outside the lower Mekong Basin and hence were excluded from mapping, including the eastern provinces of Prey Veng and Svay Rieng, and parts of the southern provinces of Kampot, Kampong Som, Koh Kong, and Pursat. The rice growing soils have been mapped based in part on an old small scale map (1:900,000) of soils of the whole country. However, soil mapping coverage of the upland regions where soils are predominantly developed on sandstones and related siliceous formations are poorly described. In Cambodia, the Arenosols (sandy soils featuring very weak or no soil development) are mapped on only 1.6% of the land area. Sandy surface textures are more prevalent than the deep sandy soils that fit the definition for Arenosols. Sandy textured profiles are common amongst the most prevalent Soil Groups including Acrisols and Leptosols. The Acrisols are the most prevalent Soil Group occupying nearly half of the land area of Cambodia. The main subgroups are: Gleyic Acrisols (20.5%, Haplic Acrisols (13.3%), Plinthic Acrisol (8.7%) and Ferric Acrisol (6.3%).

Rice Soils

Rice is the dominant crop in Cambodia, with a production area equivalent to 90% of the agricultural land in Cambodia. The sandy Prey Khmer Soil Group comprises 11% of the rice-growing soils. The soil groups have been defined using the Cambodian Agronomic Soil Classification (CASC) which groups soils according to their effects on lowland rice production. It is adapted from the Fertility Capability Classification rather than a soil classification based on concepts of soil genesis that emphasizes subsoil properties. The Cambodian Agronomic Soil Classification emphasizes surface soil properties since rice roots are relatively shallow. Even Prey Khmer Soil Group when used for lowland rice would generally not fall into Arenosols since only the 0-50 cm layers are considered in CASC.

The Prey Khmer soil has very low CEC, organic C, total N, exchangeable K and Olsen P. In field trials in Cambodia, strong responses to N are generally reported in sandy rice soils. However none of these soils respond to N alone. On the sandy Prey Khmer soils, N alone either has no effect on yield or decreases it. On sandy soils, responses to P alone may be obtained although strongest responses generally require N and P, and on the lower fertility soils K and S fertilizers are also required for rice. Low levels of Mg and B have also been identified as potential production constraints for crops on the Prey Khmer soils, but have not been verified in rice in the field. Leaching of N and other nutrients may also limit productivity of these soils even when water is not limiting. The Prey Khmer soil in Cambodia has low potential productivity even with fertilizer application.

The dominant rice ecosystem in Cambodia is rainfed lowlands. The shallow, drought- and submergence-prone sub-ecosystem, is most widespread of the rice sub-ecosystems in Cambodia, in part due to the erratic rainfall, topography and the prevalence of sandy textures in the root zone of the rice crop. While the sub-ecosystem concept is useful in regional classifications of rice growing areas, in practice local surface hydrology can vary to such an extent as to overrides the influence of rainfall. Within a single farm or among adjacent fields, the upper terraces which are commonly sandy may be classified into the drought-prone sub-ecosystem and the lower terraces may belong to the submergence-prone or drought- and submergence-prone sub-ecosystem. Fields in the high or upper terraces of the lowlands lose large amounts of water, particularly after heavy rainfall, through surface runoff and subsurface lateral water movement, while those in the lower terraces may intercept the flows from the upper paddies. Location of on-farm drains, road embankments and drains under roads can markedly affect where the runoff is directed. Water balance models are particularly useful for identifying key aspects of the surface hydrology experienced by rainfed rice. Fukai *et al.* (1995) have developed a water balance model for sandy soils in N.E. Thailand and this model may be useful for the sandy soils of Cambodia. Maintaining water in the root zone for rice is hindered on sandy soils by high percolation rates that are a common problem in the sandy lowland rice soils of Cambodia.

In the rainfed lowlands, significant periods of loss of soil-water saturation occur intermittently throughout the growing season. Based on rainfall, its distribution and variability, it could be assumed that drought was the main soil water-related constraint for rice in the region. However, the more common effect of low soil water may be to limit nutrient availability and uptake rather than to cause drought per se. The implications of the temporary periods of loss of soil-water saturation for nutrient availability are not fully understood , although variations in soil water saturation interact with nutrient availability. Fluctuating soil water regimes will have major effects on the forms and availability of N , P and on Fe and Al toxicities.

Options for minimizing the impact of periods of loss of soil-water saturation are either to use cultivars that are efficient in P uptake and use, and presumably would be best able to cope with a temporary decline in P availability ; or to treat soil with straw. Straw keeps the redox potential lower during the period of soil-water saturation loss, thus decreasing the extent of $Fe2+$ oxidation and minimizing losses in P availability due to reaction with Fe oxides. Other forms of organic matter added to the soil at planting, including cow manure, or residues from pre-rice pulse crops or green manures like sesbania, can all help minimize losses of P during periods of soil-water saturation loss.

Iron toxicity has been reported for Prey Khmer soils in Cambodia. However, the impact on yield has not been quantified. Neither is there direct evidence of the consequences of intermittent loss of soil water saturation on the incidence and severity of Fe toxicity.

Application of clay to sandy soils has been suggested as a semi-permanent treatment to enhance water and nutrient retention. Initial research on the sandy soils of N.E. Thailand suggests very strong responses in growth can be achieved by clay amelioration. The use of claying presumes a ready local supply of clay. N.E. Thailand has numerous deposits of high activity clay in lacustrine sediments (S. Ruaysoongnern, personal communication). The relevance of this technology for the Prey Khmer (Arenosols) of Cambodia, warrants further research.

Upland Sandy Soils

Important upland crops in Cambodia are maize, rubber, soybean, mung bean, cassava, sesame, peanut and sugarcane. There is very limited information on the sandy upland soils of Cambodia. Only generalized comments can be made at this stage, based largely on understandings developed for rice soils with similar properties and on recent studies carried out in the west of Takeo Province where sandy soils are prevalent.

The Prey Khmer soil is defined for rice production as having a sandy layer <50 cm deep, because deeper sand is unsuitable for rice. However, similar soils to the Prey Khmer are encountered in Tramkak with deeper sandy layers up to 80 cm. These soils are suitable for non-rice field crops and so the deep phases have been distinguished from the Prey Khmer as defined by White *et al.* (1997). The surface soil properties are similar to those reported above. That is, low levels of organic C, N, Olsen P, exchangeable K are commonly found in surface layers. In addition, KCl 40 extractable S levels, DTPA Cu, and Zn, and hot CaCl2 extractable B levels were low.

From preliminary analysis of a range of upland soils from Takeo, soil acidity appears to be a significant limiting factor for a range of field crops. In Prey Khmer soils in uplands of western Takeo Province, Al saturation values of 50-80% were found in the subsoil. Aluminum saturation >20% is commonly regarded as a potential Al toxicity in sensitive crops, whereas in very tolerant crops >80% Al saturation is required to impair crop growth. Seng *et al.* (2004a) showed strong responses by upland rice to lime application on the acid Prateah Lang soils (pH CaCl2 4; Al saturation 80%) when maintained in an aerated state whereas no response was found when these soils were flooded.

Hydraulic Properties

Water Retention Properties

Sandy soils retain little water at high water potentials and water content decreases rapidly with the water potential. Panayiotopoulos and Mullins (1985) studied the water retention properties of pure sand materials varying in size. They showed that most water was released between -0.1 and -1kPa for a coarse sand (2,000-710 μm) and between -15 and -30 kPa for a very fine sand (125-45 μm). The limited water release observed for the very fine sand between saturation and -0.5 kPa was not discussed by Panayiotopoulos and Mullins (1985). Mullins and Panayiotopoulos (1984) showed that the water retention curve was only very slightly affected by the clay content for a clay content <20%.

The clay used was a kaolinite. With sandy soils, two thirds of the water present at saturation is usually released at -30 kPa as recorded by Obi and Ebo (1995) in a sandy soil in Southern Nigeria. Water contents ranging from 0.20 to 0.30cm ^{3}cm $^{-3}$ and from 0.04 to 0.12cm ^{3}cm $^{-3}$ are often recorded at -33 and -1500kPa, respectively in tropical soils belonging to the sand, loamy sand and sandy loam textural class.

In sandy soils, there is very little water available at matric potential <-100 kPa. Kukal and Aggarwal (2004) measured a water content of 0.16 and 0.10 cm^3 cm^{-3} at -33 and -1500kPa, respectively in a sandy loam topsoil (% clay = 10%) in India. The water content significantly increased with a slight increase in the clay content and was 0.22 and 0.13 cm^{3}cm $^{-3}$ at -33 and -1500kPa, respectively when the clay content was 14%. Osunbitam *et al.* (2005) showed in Nigeria an averaged water loss in sandy soils of 0.006mm $^{-3}$ between -100 and -150kPa, the water content at -150 kPa being 0.017mm $^{-3}$. Several studies have shown that the available water increases with the silt content.

Tomasella and Hodnett (1998) compared the measured volumetric water content at different matric potentials and those estimated with the pedotransfer functions (PTFs) developed by Rawls *et al.* (1992) from the USDA soil data base. They showed that these PTFs greatly overestimate the volumetric water content when applied to sandy soils of Brazilian Amazonia. The available water capacity measured between -5 and -1,500 kPa by Nyamangara*et al.* (2001) in Zimbabwe for topsoils with a sand content close to 90% ranged from 0.159 to 0.174 m^{3}m $^{-3}$ according to cattle manure management options.

Hydraulic Conductivity

The saturated hydraulic conductivity (K_s) of sandy soils in the tropics varies within a range of values covering several orders of magnitude ($10^{-7} < K_s < 10^{-3}$ ms $^{-1}$). In a Brazilian sandy soil with very low clay contents (average content in the whole profile of 0.25%), Prevedello *et al.* (1995) measured $1.1 \times 10^{-6} < K_s < 7.5 \times 10^{-5}$ m s^{-1}. Contrasting this in another Brazilian sandy soil with only a slight change in clay content (average content in the whole profile of 6%), Faria and Caramoni (1986) measured $1.5 \times 10^{-5} < K_s < 2.8 \times 10^{-4}$ m s^{-1}.

In soils, K_s varies according to the development of the macroporosity. As a consequence, K_s variation is more closely related to the macroporosity development rather than to soil texture. Thus, most studies try to relate K_s to part of the macroporosity that is called effective porosity ($Ö_e$) and defined as total porosity, Ö minus the water content at -33 kPa. K_s and $Ö_e$ are related as following: with *B* and *n*, are two parameters varying with the soil characteristics. These parameters were obtained by Tomasella and Hodnett

(1997) for Brazilian tropical soils. They found log B=4.752 and n=4.536 for the soils studied by Predevello *et al.* (1995) and log B=4.758 and n=4.532 for soils studied by Faria and Caramoni (1986). These soils were also used by Tomasella and Hodnett (1997) to derive parameters of the Brook-Corey/ Mualem model for unsaturated hydraulic conductivity.

However when cultivated, the macroporosity of sandy soils is very unstable and collapses rapidly in the presence of water. Thus, the measurement of K_s becomes difficult to perform without modifying the macroporosity that have a tremendous effect on K_s. This probably explains why many studies do not report a large range of K_s variation between field experimental treatments and with time as expected. Thus Osunbitan *et al.* (2005) recorded $5.5 < K_s < 7.5 \times 10^{-5}$ ms $^{-1}$ in a topsoil under different tillage treatments. The K_s difference recorded by these authors between the different treatments (1×10^{-5} ms $^{-1}$) was small and similar to the differences recorded under the same tillage treatment over a period of 8 weeks.

Unsaturated hydraulic conductivity ($K_{\grave{e}}$) of Brazilian sandy soils was also discussed by Tomasella and Hodnett (1997).

Surface Crusting and Water Infiltration

Because of the very small inter-particle cohesion that results in a very small aggregate stability, sandy soils are highly sensitive to surface crusting, thus explaining the large number of papers in this area. Crusts protect the soil surface from wind and interrill erosion but they also favour runoff and consequently rill and gully erosion.

Two main types of structural crusts were recognised in sandy soils depending on the dominant forming process: (i) sieving crusts made of well sorted micro-layers with average infiltrability of approximately 30 mmh $^{-1}$, (ii) and packing crusts made of sand grains closely packed with average infiltrability of 10 mmh $^{-1}$.

Bielders and Baveye (1995b) studied in the laboratory the processes of structural crust formation on coarse textured soils. They proposed that the formation of clay-band in sieving structural crusts would be initiated by the displacement of micro-aggregates or other small particles from the above washed-out layer, followed by their accumulation due to mechanical straining. Erosion crusts that result from smoothening and erosion of structural crusts and depositional crusts that result from sedimentation were also described. They exhibit more restrictive infiltrability (2-5 mm $^{-}$h^{1}) than structural crusts.

The development of crusts leads to runoff that can be quite significant. Sombatpanit *et al.* (1995) measured between 300 and 400mm of runof f that corresponded to about 35% of the rainfall on bare sandy soils in Thailand with 25 to 70 tha $^{-1}$ of soil loss. Runoff was still between 10 and 20% of the rainfall under different agricultural treatments.

Surface infiltrability can also be reduced in sandy soils by repellency. Indeed, sandy soils are particularly susceptible to water repellency and susceptibility increases with the duration of the dry season. Repellency is responsible for vertical fingered flow in sandy soils because of the presence of repellent soil volumes with hydrophobic organic matter (Roberts and Crabon, 1972; Dekker and Ritsema, 1994; Ritsema and Dekker, 1994). Study of repellent soils in arid and humid climates showed that repellency would be much more related to the type of organic matter than to the duration of the dry period.

Compaction

Sensitivity to Soil Compaction

Unlike other soils, the structure of sandy soils can be easily affected by mechanical compaction over a large range of scales. Usually mechanical compaction preferentially affects large pores (*i.e.* macropores that result from tillage and biological activity) but in sandy soils it affects these large pores down to the small pores that result from the arrangement of the skeleton particles (sand and silt) within the clay fraction. That re-arrangement when submitted to mechanical compaction is possible because of the small cohesion between the skeleton particles.

For narrowly graded pure sand materials, Panayiotopoulos and Mullins (1985) showed that these air-dry and nearly saturated sands were always found to pack more closely under a given load than the same sand at any water content.

Very small porosity can be recorded under wheel tracks and just underneath the tilled layer. Thus, Bennie and Botha (1986) recorded $1.7 < D_b < 1.8 gcm^{-3}$ in the 0-20 cm layer under wheel tracks and in the 20-40 cm layer. Because of this small inter-particle cohesion, high bulk density is also recorded when sandy soils are puddled in rice-wheat cropping systems. Thus, Aggarwal *et al.* (1995) recorded $1.75 < D_b < 1.82 gcm^{-3}$ in the 15-20 cm layer after several years of high puddling in a sandy loam soil. Smith *et al.* (1997a) studied the effect of soil compaction on a large range of South African forestry soils. They showed on soil cores in the laboratory that the porosity after compaction of sandy loam and loamy sand soils was related to the size distribution of the sand fraction and tended to decrease with the increase in the clay and silt content.

Smith *et al.* (1997b) showed for loamy sand that increases in compaction were almost independent of the water content and then almost entirely due to increasing applied pressure alone. Smith *et al.* (1997c) also recorded a high compactibility for sandy soils derived from sandstone, granite and aeolian sands. The maximum bulk density was related to the loss in mass after ignition at 450°C.

Penetration Resistance

Increases in bulk density invariably results in an increase in the penetration resistance with significant consequences for root development although there no clear relationship with the penetration resistance (Mullins *et al.*, 1987; Bengough and Mullins, 1991). Critical values that severely restrict root growth have been estimated to vary from <1 to >4 MPa depending on the soil, water content and crop type. Indeed, the penetration resistance varies within a large range of values according to the soil water content without any variation of the other soil characteristics. It is significantly inversely related to water content. Many penetration resistances published in the literature for sandy soils range between 0.1 and 0.8 MPa but the water content at which they were determined often remains unclear. Bruand *et al.* (2005) recorded a penetration resistance ranging from 0.35 to 0.55MPa in the subsoil of a sandy soil in Thailand when the water content ranged from 0.03 to 0.09kgkg $^{-1}$. In South African sandy soils that developed in aeolian sand, Du Preez *et al.* (1981) measured penetration resistance >1.5MPa, at field moisture capacity, in a ploughed layer under wheel tracks.

In these soils Bennie and Botha (1986) confirmed the presence of such values of resistance to penetration in the subsoil and showed that they result from compaction because of traffic that leads to an increase in the penetration resistance that restricts root development for wheat and maize. Kukal and Aggarwal (2003) measured much greater penetration resistance in a sandy loam soil after puddling because of subsurface compaction. Indeed, these authors recorded at field capacity a penetration resistance ranging from 3.0 to 4.5 MPa in the compacted layer.

In their study, Smith *et al.* (1997a) showed on compressed soil cores only small differences in strength development across a wide range of water content for a loamy sand soil. This would be related to the contribution of frictional rather than cohesion forces to resistance to penetratio. On the other hand, results also showed somewhat different behaviour for a sandy loam soil, resistance to penetration increasing from 1 to 5 MPa over a range of water content of only 4% by mass.

This large range of resistance to penetration would result from the contribution of cohesion forces that are partly related to the water content. Thus, a decrease in water content would increase frictional and cohesion forces from field capacity to intermediate water content, smaller water contents increasing the frictional forces alone, the cohesion forces disappearing, thus explaining the results recorded in Nigeria by Ley *et al.* (1995) on a large range of soils including sandy soils.

Effect of combined deep tillage and controlled traffic on penetration resistance and its consequences for root growth has been studied in several countries. In their one-year study, Bennie and Botha (1986) showed that deep ripping and controlled traffic led to a significant increase in rooting depth, rooting density in the subsoil, water use efficiency and yield increases of 30% for maize and 19% for wheat.

Increased yields were recorded in many earlier studies after deep tillage but the duration of the positive effects of deep tillage is still under discussion. Slotting was also used to loosen the subsoil in sandy soils of Northeast Thailand and thus promoted rooting.

Hartmann *et al.* (1999) showed that rooting depth and yield of various crops were increased for two successive years after slotting. Lesturgez (2005) recorded a significant increase in root density in the slot that enables a better subsoil exploitation.

Lesturgez *et al.* (2004) also investigated the potential use of forage legume *Stylosanthes hamata* (stylo) to ameliorate the structure of compact layers in sandy soils of Northeast Thailand. They showed that after 24 months of continuous stylo, roots were able to penetrate the compact subsoil, resulting in an improvement of its macroporosity. They also showed that a subsequent maize crop developed a deep and extensive root system using the macropores.

Hardsetting in Sandy Soils

Many tropical sandy soils are potentially hardsetting soils, *i.e.* they can become compact, hard with apparently apedal condition prevailing on drying. In these soils, a significant increase in soil strength is recorded over very narrow water content changes within the plant available range of soil water potential (-10^2 to -10^3kPa) with resulting adverse effects on root growth and crop production.

Thus, Chan (1995) measured strength characteristics in sandy loam hardsetting soil in the semi-arid region of Australia and showed that in the cultivated soil strength increased from 0.02 to 0.09 MPa with decreasing water content from 0.11 to 0.04 kgkg $^{-1}$ when there was no strength variation under permanent pasture. McKyes *et al.* (1994) studied the cohesion and friction in two sandy-loam hardsetting soils from Zimbabwe. They showed that the cohesion changed from nearly zero at saturation to well over 0.1 MPa in the field dry state.

Young and Mullins (1991) suggested that the amount of $<60\ \mu$m particles rather than solely the $<2\ \mu$m is important in causing the development of hardsetting properties.

High soil strength in sandy soils can be also partly related to the development silica precipitation that forms globules and silica flowers over the sand grains. These precipitations would not be responsible for a cementation of the sand and silt grains but would lead to an increase in particle contacts and frictions, thus explaining the increase in soil strength recorded.

Controlled Compaction

Compaction in sandy soils was also discussed as a possible water and nutrient management to improve water retention properties and reduce nutrient leaching in Indian sandy soils. Indeed, according to these authors, compaction that reduces the volume and continuity of large pores, would increase water retention and reduce water infiltration and saturated hydraulic conductivity in highly permeable deep sandy soils. Compaction would save irrigation water by 15-36% and increase productivity by 30-50%.

Puddling is also used to reduce high percolation losses of irrigation water and nutrient leaching when cultivated for rice production. Sharma and Bhagat (1993) showed that puddling was effective in reducing percolation losses when sand was less than 70%, and finer fractions were dominated by clay (13-20%). Puddling has been reported to decrease saturated hydraulic conductivity of the puddled layer (0-10cm) of sandy loam soil from 1.810 $^{-7}$ ms $^{-1}$ in unpuddled to 4.210 $^{-8}$ms $^{-1}$ with medium puddling and 2.510 $^{-8}$ms $^{-1}$ with highly puddle soils. Kukal and Aggarwal (2003b) showed in a sandy loam soil that puddling reduced percolation losses by 14-16% with the increase in puddling intensity from medium to high, whereas the amount of irrigation water required decreased by 15-25%. Similar results were recorded by Kukal and Sidhu (2004) in another sandy loam soil in India.

However, puddling in rice-wheat cropping systems leads to some adverse effects for the following wheat crop that requires to be managed with appropriate tillage techniques.

A yield decline of wheat is often recorded because of subsurface compaction ($1.70 < D_b < 1.75$ gcm $^{-3}$) at 14-20 cm depth under normal puddling at normal depth. Kukal and Aggarwal (2003a) showed in India for a sandy loam soil that puddling at shallow depth (5-6 cm) led to the development of a compact layer at 10-12 cm depth that was loosened ($1.50 < D_b < 1.55$ g cm^{-3}) during normal cultivation for wheat seedbed preparation

In the tropics, physical attributes of sandy soils are particularly sensitive to both the sand and silt size distribution and mineralogy of the clay fraction. Because of the presence of low activity clay in most sandy soils, the assemblage of elementary skeleton particles is highly unstable resulting in a high instability with respect to structure from the microscopic to macroscopic scale. When for a variety reasons, the assemblage collapses, the resulting porosity and penetration resistance would be all the greater as the skeleton particle distribution is heterometric.

In contrast, in sandy soils unlike other soils, the elementary fabric can be easily loosened by tillage practices. Thus greater porosity can be produced easily by tillage but its stability is very weak and compaction by wheels or other actions can produce a dense structure with adverse physical properties. This leads to a decrease in the water retention properties and hydraulic conductivity, an increase in the resistance to penetration and sensitivity to surface crusting.

More generally, tropical sandy soils, more than other soils, require careful management in an environmentally friendly manner. Indeed, even if most physical degradation processes are more easily reversible in tropical sandy soils than in other soils, the physical fertility of these soils is weak. These soils require very little tillage operations in the wrong way to produce significant adverse consequences for plant development and consequently for crop yield and environment.

Compaction Processes in a Tilled Sandy Soil

Soil compaction in agricultural systems is a worldwide concern and has received considerable attention over the past decades. Soil compaction is defined as: "the process by which soil grains are rearranged to decrease void space and bring them into closer contact with one another, thereby increasing the bulk density".

The vast majority of soil compaction in modern agriculture is often attributed to heavy machinery and traffic load. However other processes can be involved and soil compaction may occur without traffic load on soil surface. For example, the formation of a dense subsoil layer known as "fragipan" is interpreted by soil collapse under its own weight. This process occurs when a metastable arrangement of particles is wetted under a constant confining pressure (weight of the top layer in the case of natural collapse).

Sandy soils are often considered as structurally inert because of their weak structure and the absence of shrink-swell properties but frequent and severe compaction observed in agricultural fields raises the question of the processes and factors that control soil compaction and its reversibility. In the sandy upland soils of Northeast Thailand, subsoil compaction (20-40 cm) is a common feature that impairs root development and therefore is responsible for low crop production.

Comparisons between forest and adjacent cultivated area have proved that the compact layer commonly observed in agricultural fields was induced by intensive agriculture of the last decades. Sandy soils of the region have developed mainly from light textured aeolian material well known for its problematic characteristics for engineering works. On the other hand, deep ploughing and subsoiling have always been inefficient in overcoming this compaction in agricultural systems since soil re-compacts after the first heavy rain. This suggest that collapsibility is a key factor not only in deep profiles but also in the superficial and tilled layers.

Compaction of aggregates beds in a dry state, hydrocollapse (also known as hydroconsolidation) and load-upload cycles under the same pressure (traffic load conditions), can be involved in the formation or the reformation of a compact layer. The objective of the study was to investigate the processes of soil compaction in a tilled sandy soil subjected to non-flooding rains and evaluate their respective contribution to total soil compaction. Experiments focused on uniaxial compactability, hydrocollapse and rearrangement under traffic load.

Soil Characteristics and Sampling

The samples were collected in a sugarcane field located in Ban Phai District, 40-km from Khon Kaen City, Northeast Thailand (16°082 N, 102°442 E). The choice of the site was based on a previous investigation that highlighted the presence of a compact layer located at 20-40 cm depth that was representative of the general situation of subsoil compaction. The soil has a sandy texture with no or very weak structure. It belongs to the Nam Phong soil series and was classified as a loamy, siliceous, isohyperthermic Arenic Haplustalf or Arenic Acrisol. Three undisturbed samples were collected from the vertical face of a pit, respectively in the topsoil (0-15 cm), in the compact subsoil (15-25 cm) and underneath the compact layer (40-50cm). Mineralogical characteristics of the studied soil were investigated using X-ray diffraction. When the sand and silt fractions were exclusively constituted of quartz, the clay fraction included kaolinite, traces of illite and a significant proportion of small quartz particles. The three samples were identical in their mineralogy and the particle size distribution of the sand fraction. They differed only in their clay content (from 70gkg $^{-1}$ in the topsoil to 136 g kg^{-1} in the deepest layer). A last sample (pure sand material) was prepared from the topsoil horizon by sieving the >50 μm material from the whole soil after dispersion in sodium hexametaphosphate.

Sample Preparation

The samples were manually crumbled in the laboratory in order to produce small aggregates similar to tillage-induced aggregates. The aggregates were poured into a ring 50mm in diameter, and 18mm in height placed on a porous plate using a small funnel fixed 5-cm above the middle of the ring. The ring was overfilled, then the surface was carefully levelled off, and the assemblage thoroughly cleaned with a small brush. The assemblage was then installed in the oedometer and the top cap gently positioned. Preliminary tests had shown that the preparation of the aggregate beds using this method allowed the formation of a metastable arrangement of aggregates with a bulk density similar to that of the topsoil after ploughing.

Design of the Oedometer Apparatus

The oedometer test is classically used for consolidation and compression studies of fine-grained soil samples, such as clays and silts, since it recreates the conditions of volume change with zero lateral strain (*i.e.* one-dimensional compression). The oedometer apparatus used allows the application of an axial load that ranges from 0 to 1,500 kPa. We applied a range of 29 pressure steps (2, 5, 10, 15, 20, 25, 30, 35, 40, 45, 50, 60, 70, 80, 90, 100, 150, 200, 300, 400, 500, 600, 700, 800, 900, 1,000, 1,100, 1,300 and 1,500 kPa), with 5 minutes interval between each step (minimum duration to reach equilibrium).

The change in volume of the samples was recorded continuously by measuring the vertical displacement of the rigid top platen used to apply the load. The design of the apparatus allows the injection of water on the top of the sample. Drainage was free through the porous plate located below the sample.

The volume of water injected into and drained from the samples can also be recorded. Bulk density and average water content were derived from these measurements.

The background noise of the oedometer originating from internal deformation (porous plates) and elasticity of the membrane was estimated during preliminary tests on uncompressible Plexiglas cylinders and results were corrected accordingly.

We used the following experiment to characterize the compaction in dry and wet conditions:

a. To characterize the behaviour in dry conditions, an air-dried aggregate bed of the control and the three horizons was prepared with five replicates. The series of pressure steps was applied and the changes in bulk density recorded.
b. To characterize the behaviour in a wet state, an identical set of aggregate beds was prepared. The beds were saturated by injecting water under no load until drainage began at the bottom of the samples. The series of pressure steps was then applied while the samples were kept saturated in a free draining state. The changes in bulk density were continuously recorded.

Hydroconsolidation Tests

Hydroconsolidation is characterised by an abrupt change in bulk density of samples loaded at their *in situ* water content and then flooded. Hydroconsolidation under a constant load P_w was studied in a three-stage test:

a. Air-dried samples (5 repetitions for each depth) were loaded step by step to a constant load P_w of 2, 100, 500 and 1,500kPa.
b. While the axial load P_w was maintained on the sample, water was injected at 50mm ^{3}min $^{-1}$ through the porous plate located on the top of the sample until drainage started at the bottom of the porous plate.
c. The load was then increased step by step on the wet samples from P_w to 1,500kPa. Water continued to be added freely at no pressure to keep the sample wet until the end of the test.

Compression-relaxation Tests

The purpose of this test was to characterise subsequent compaction (*i.e.* rearrangement processes) under a series of identical axial loads in wet conditions. The test consisted of a series of compressions/ relaxations applied on wet samples:

a. Five samples of each horizon were saturated and then loaded step by step to create stresses P_R of 100, 500 and 1,500kPa. Water was added freely at no pressure to keep the samples wet throughout the test.
b. A series of 70 cycles of compression (P_R) and relaxation (0 kPa) was applied on the wet samples. Bulk density was continuously recorded.

Results

The compaction curves using the classical compaction approach. Both dry and wet bulk density measurements are presented as a function of axial load. For the pure sand, compaction due to axial load was very low over the range of pressures and there was no significant difference between dry and wet curve.

Compaction was low and highly heterogeneous between replicates for the three soil samples up to 25kPa. There was no significant difference between the dry and wet samples in this range of pressures. Bulk density increased sharply above 25kPa and became more homogeneous. Beyond 100 kPa, the bulk density increased with depth for any given load all soil samples. At 1,500kPa the dry bulk densities reached 1.60, 1.63 and 1.69 $Mg.m^{-3}$ for the 0-15, 15-25 and 40-50cm depths, respectively.

The results of the hydro-consolidation test on the pure sand material. The compaction either in the dry or wet state was almost insignificant and there was no significant difference between the dry curve and the wet curve at any pressure. Therefore, the collapse was insignificant. The results of the hydroconsolidation test together with the results of the compaction test for the 25-35cm depth layer. Collapse resulting from water injection (represented on the chart as white arrows), always resulted in a final bulk density between the dry and the wet compaction values. The increase in bulk density as a result of hydrocollapse was quite similar over the range $50 < Pw < 1{,}500$kPa, even though the largest bulk density change was recorded for an axial load of 100kPa. The bulk density after hydrocollapse was sometimes significantly lower than that of the wet curve ($P < 0.05$). However, the difference was no more significant when the loads were increased after hydrocollapse had occurred. As the dry and wet compaction curves tend to get closer at high pressure, we can assume that for very high loads, no more collapse will occur. The changes in bulk density with increasing water content at constant load (Pw= 1,500 kPa) for the three soil horizons. Two replicates are presented for each depth as a means to highlight heterogeneity. In agreement with the compaction in wet conditions, the lowest collapse was recorded for the 0-15cm depth sample, and the intensity of collapse increased with depth. Collapse always occurred in a range of gravimetric soil moisture between 5 and 15%.

The compression-relaxation curves. The bulk density increased by at least 0.1T.m^{-3} at P_R=1,500kPa from the initial compaction to the end of rearrangement cycles. The soil samples presented significant elasticity. The material recovered a significant proportion of the porosity when the axial pressure was relaxed but some of the deformation is non-reversible and the bulk density increased for each cycle. The deformation intensity decreased as the number of cycles increased. The rearrangement intensity increased with depth. Similar results were obtain at P_R=100 and 500kPa (data not shown).

Compaction of the Pure Sand

For the pure sand, the sensitivity to compaction, in dry and in wet conditions as well, was very small and can be considered as being independent of the applied pressure. The sand grains did not reorganize under pressure, even when wet. This result suggests that the lubricant effect of the water was ineffective in the case of this material. Two factors can explain this unusual behaviour. Firstly, the bulk density was already 1.46T m $^{-3}$ at the beginning of the experiment, probably because the size of the sand grains was distributed over a large range. Secondly, most sand grains had a jagged shape according to them aeolian origin, that probably resulted in inter- locking between the grains.

Compaction of Soil Samples

In contrast, the compaction curves recorded with the soil samples in wet condition proved that the same sandy material mixed together with clay and silt was highly compactable. In dry conditions compaction started at around 25kPa and increased with increasing pressure until 1,500kPa. In the case of aggregate beds, collapse was in part the consequence of the deformation of the aggregates. This process was not active in the pure sand material under study because aggregates were absent. Dry compaction may in part result from the deformation of clay particles. However, the contribution of this process must be limited, given the low clay content of the material and the high proportion of quartz grains within the clay fraction. The major contributing factor associated with compaction was probably due to lubrication, the planar-shaped clay minerals helping the sand grains slip against each other.

Hydrocollapse

When water is injected in the samples, hydrocollapse proved to be a phenomenon that developed fully under constant pressure and at any given pressure. Indeed, whatever the initial pressure, the final bulk density was almost identical to the bulk density obtained by compaction in wet conditions. This result, consistent with the observations of Assallay *et al.* (1998) on loess materials, has a direct application in predicting the collapse. Maximum collapse under any load can indeed be estimated by the difference between the dry and wet compaction curves under the considered load. Maximum hydrocollapse was recorded for P_w=200kPa, close to the value of 100kPa observed by Assouline *et al.* (1997) on aggregate beds. It has been shown in aeolian deposits that collapse needed a small amount of clay to develop , and that collapse intensity increased with clay content up to 25% clay. The same increase in hydrocollapse with clay content was observed in this experiment, but the range of clay content covered by the three samples was not sufficient to determine a maximum value.

The increase in water content with clay content for hydrocollapse to develop suggests that the process is related to the hydration of the clay minerals. Faure (1976) mentioned the importance of the clay fraction in compaction of sandy soil and introduced the notion of water potential and clay hydration. In the three horizons hydrocollapse started between 3 and 7% of gravimetric water content. This low water content proves that the phenomenon becomes active in any horizon as soon as it gets wet. The samples presented a mechanical behaviour similar to metastable deposits. These properties, usually associated with loess and loess-like deposits, are therefore not confined to silty materials and develop also in sandy soils.

Compression-relaxation Cycles

The mathematical description of soil compaction is based on relationships between bulk density and applied stress. This approach assumes that after a sample has been consolidated under a pressure P_1, the consolidation would resume only for a pressure $P_2 > P_1$. This theory is not applicable to the results of this study as a series of successive stresses under the same axial load resulted in a substantial increase in bulk density. The relaxation between successive stresses allowed the internal friction between sand grains to decrease and therefore permitted the network of forces to reorganise during the next axial load, leading to increased bulk density.

The asymptotic shape of the curve showed the development of the soil structure towards the highest possible bulk density. The rearrangement test in a wet state is probably the most representative test to simulate vehicle traffic load as it models as series of confined uniaxial stresses under the same pressure.

Contribution of the Different Processes to Total Soil Compaction

The contribution of dry compaction, hydrocollapse and compression-relaxation cycles to bulk density increase was estimated from our results at a load of 1,500kPa. The last series, namely "field" is the bulk density measured in the field using cylinders. The effect of the three processes on bulk density increased with clay content. However, the contribution of the three processes to bulk density increase remained similar in relative value whatever the clay content.

Dry compaction represented around 50% of total compaction, when hydrocollapse and rearrangement ranged between 20 and 30%. In the field dry compaction and hydrocollapse under low pressure (weight of the upper soil horizons) are the first two processes to develop after tillage. The many tillage operations usual in the region induce through a succession of traffic loads, rearranges the fabric to produce the usual massive aspect of the soil with high bulk density. The close lay out of grains, with small particles filling the voids left between bigger ones, has been described by Bruand *et al.* (2004) as the main factor of high resistance to penetration of the compact layer.

Finally, as soil sensitivity to compaction increases with clay content and clay content increases with depth, the most sensitive horizons are the deepest. As a consequence the deeper the soil tillage, the higher is the risk of compaction and the final density. The bulk density is highest in the 20-40cm layer probably because this layer supports the wheels of the tractors during ploughing (at least three times a year). Surface axial load due to vehicle traffic may also be transmitted to subsoil horizons through the massive and often dry topsoil, increasing bulk density through rearrangement processes. In the field the highest bulk density values were recorded in the 20-40 cm depth interval despite the higher sensitivity to compaction of the lower horizon. Two kinds of tillage operations are to distinguish: frequent tillage of the 0-20 cm interval depth and some punctual deep tillage operations with the objective to break the compact layer.

As the soil is highly collapsible and sensitive to rearrangement, the density of the post-tilled layer reach inevitably high values as a function of time. The frequency of tillage in the topsoil (0-20cm) does not allow high density as in the 20-40cm interval depth as the structure return frequently to the tilled state.

However, the 20-40cm interval depth benefits as it has the enough time to accumulate the combine effect of the traffic load. As for the lower layers (>40 cm depth), tillage operation have never change organisation of the structure and if aggregates beds from this layer are highly sensitive, the actual weakly developed structure is stable. It has been shown that tilled layers are much more sensitive to compaction than massively structured or already compacted subsoil. Porosity of this layer is mainly constituted of biopores which are usually stable because they develop in a stable structure. These results suggest that the deeper the soil is tilled, the higher is the risk in obtaining high bulk densities. Deep tillage is therefore not an option to rehabilitate compact subsoils due to the instability of the resulting structure. However, alternative techniques such as slotting or biological drilling have proved to be efficient is such unstable soils. These techniques ensure the development of pathway for the roots through the compact layer and preserve the massive and stable structure surrounding them.

The compaction of the sandy material studied under uniaxial load was trivial, even in wet conditions. The same material was highly sensitive to compaction when mixed with silt and clay. The sensitivity to compaction increased with increasing clay content. Compaction in the dry state, hydrocollapse (collapse under increasing water content at constant pressure) and rearrangement under a series of successive loads were more pronounced when clay content increased. However, the contribution of each phenomenon to final bulk density was approximately constant whatever the clay content. Most part of soil compaction (around 50%) was due to dry compaction. Hydrocollapse explained about half of the remaining compaction. Hydrocollapse was responsible for sharp increases in bulk density as a result of small increases in water content (gravimetric water content between 3 and 7%), even under low pressure.

The rearrangement under successive loads explained 20 to 30% of the final bulk density, even though the bulk density was already higher than 1.65Mgm $^{-3}$ after dry compaction and hydrocollapse. As clay content increased with depth, the deeper horizons were the most sensitive to compaction. The highest bulk densities were however measured in the 20-40cm layer in the field. The direct traffic load resulting from the many ploughings a year usual in the region is probably a part of the explanation but the structural effect of deep ploughing (which changed the massive structure of the layer into a metastable organisation of aggregates very sensitive to densification) is probably the main factor. Deep tillage is therefore not an option to rehabilitate compact subsoils due to the instability of the resulting structure and alternative techniques conserving part of the initial stability are recommended.

Physical Reorganization of Sand Due to the Motion of a Solid Intruder

Sandy soils are very unstable or evolving materials, whose properties are not well known, making it thus difficult to manage for agricultural purposes. Due to the low content of clay, the cohesion between grains of the soil is low and the sandy soil can be particularly sensitive to any external perturbation, whatever the origins are: climatic (intense rain, capillary rise) or anthropogenic (ploughing, mechanical vibrations) for example. In this work, we are considering the sandy soil as an example of fragile matter from a physical point of view; our approaches are to identify the most simple elementary mechanisms responsible for the instability and the reorganization of the sandy structure and thus to characterize the physical parameters involved in the weakness of this type of soil. We thus use a very simple model of granular soil without any biological or chemical influences. This approach aims to test the part of the physical influence in the instability of real soils.

We will describe how we reduce the problem to model physical experiments and mention the characteristic features of dry granular media. We will describe in more details a particular experiment for testing the instability of a granular media. In particular the resistance to reorganisation of a granular medium is characterised by the ability of an intruder to move in it. The strength needed to displace it and the size of the reorganisation around critically depends on the original packing of grains and could give information on the ability of a worm or a root to penetrate this type of medium.

Model granular media for experiments: Granular matter we use in physics are collections of grains of controlled dispersity, simple form and known grain density (Fontainebleau sand made of rounded grains of 300 μm of diameters, monodisperse glass beads or metallic ball-bearing beads...). Their diameters are always larger than the micrometer, so that the thermal agitation does not play any role in the motion of the grains. Moreover, we choose large diameters (of the order of 100 μm to several millimetric sizes) to minimize for example drag forces produced by the interstitial fluid (generally air in the porosity of the granular network) so that the interactions are mediated through the direct contacts between grains. In the case of non-cohesive grains, these interactions are limited to collisions and contact forces, according to the duration of contact between grains.

Despite these drastic simplifications which seem to be very naive from a practical point of view, the variety of phenomena observed with model granular media is very rich and still complex. We will describe below the main features of a granular structure which could play a role in a real sandy soil.

The Heterogeneity of Forces

In granular materials, force is rarely transmitted uniformly, but rather preferentially along a network forming force chains. The network of contact forces can be observed by photoelastic measurements. It appears to be very heterogeneous, forming chains along which the forces are particularly intense encompassing regions bearing relatively small loads. Then the description of the transmission of forces inside a granular medium is quite a great challenge where the inhomogeneity of the material leads to unexpected behaviours.

The spatial distribution of forces is large, with fluctuations of the order of the mean force. According to the network of contact forces can be divided into 2 parts: the large network for forces larger than the mean force bears most of the load with a stress tensor and angular distribution of contact directions which is suited to oppose external constraints. On the other hand, the small force network (forces smaller than the mean force) shows isotropic contact direction and a weak anisotropy of the stress tensor with the major principal axis oriented in such a way that it opposes the buckling of large force chains.

These chains play an important role in many of the properties of the granular material, such as the transmission of sound and the fragility of the packing along particular directions. Generally the mesh size of the large force network is of the order of 10 grains, which is also the typical size of a shear band.

The History Dependence

The protocol of preparation is particularly important for granular materials and determines the subsequent mechanical properties. For example preparing a dry sand pile by using a point source (sand falling from a funnel) or a pluviation technique (sand falling from a grid) leads to the same macroscopic pile with a repose angle (angle between the slope of the pile and the horizontal) which is around 30° for both piles.

However the measured pressure profiles on the bottom below the sand piles are rather different. When the sand pile is built from a point source, there is a dip of pressure below the apex of the pile, while there is a maximum of pressure when the pile is prepared by pluviation. This drastic difference in the pressure profiles can not be inferred from macroscopic properties like the angle of repose and therefore the internal coefficient of friction is not sufficient for describing the mechanical properties of a sand pile. Some microscopic parameters have to be taken into account.

The importance of preparation can act on packing fraction but also on finer parameters like the directions of contacts between grains and the orientations of the force network. The evolving nature of the mechanical properties of a granular medium can be illustrated by the following experiment where some photoelastic discs are placed into a shear box.

Initially before applying the shear the medium is isotropic (there is no anisotropy in the orientations of contact directions) and the response to a point force on top of the box is maximal along the direction of the force (applying a point force is a way to test the mechanical properties). However if the box is submitted to an nitial preshear some new force chains are created along direction 1 which tend to oppose the shear. A new "texture" is formed and the mechanical properties have changed, as it can be observed on the shift of the response to a point force.

The Fragility

Granular materials as other particulate materials are examples of fragile matter. For non-cohesive grains there is a lack of resistance of contact forces to any extension. The internal structure (the contact and force networks) can evolve and adapts itself to support the applied load as we have seen before.

Then the incremental response can be elastic only to "compatible" loads. Incompatible loads like the one produced for example by a change of compression axis, even if small, will cause finite, plastic reorganizations: irreversible rearrangements will be produced in the structure.

The Different Scales

The approach used at a macroscopic scale in soil mechanics is usually based on standard compression experiments like triaxial tests, from which constitutive relations between stresses, deformations and directions of deformations are obtained. In this approach, the soil can be considered as an effective continuum medium but the constitutive elasto-plastic relations between stresses and strains inferred from these tests are mostly non-linear, piece-wise, anisotropic and crucially depend on the history of the loading/unloading cycles.

The problem of granular materials is that there is no clear separation between microscopic and macroscopic scales, from the size of the micrometric asperities in the surface area of contact of grains to the mesoscopic scale of force chains and till the macroscopic size of the bulk material of soil. Therefore many scales are involved in the resulting mechanical properties and microscopic rearrangements can have a drastic effect on macroscopic properties. This is the reason why we perform the following experiment.

Experiment on reorganizations in a granular medium due to a solid intrusion: In this part we describe in more details a conceptually simple experiment. It consists in testing the local resistance of a granular material by moving an intruder in it.

From a practical point of view it bears similarities with standard penetrometry tests currently used in soil mechanics. However we want to focus here on the microscopic perturbation introduced by the measurement itself, which is a consequence of the fragile nature of granular material. The network of contacts between grains is not permanent and the perturbation induced by the motion of an intruder (which can be a rough model for the growth of a root or the progression of a worm) can open or close some contacts and produce some irreversible rearrangements that change the nature of the granular structure itself and thus of the complementary porous matrix, what affects its mechanical properties at a larger scale.

Therefore we want to characterize the rearrangements induced by the displacement of the intruder and the range of the effect of perturbation inside the granular material. Thus we apply a local and cyclic perturbation inside the granular packing for both detecting the displacements of grains in the vicinity of the perturbation and characterizing the evolution of the structure by investigating the irreversible displacements after a given number of cycles of perturbation.

Experimental Setup

For this purpose we use a 2 dimensional (2D) granular media. That means the motions of grains can only occur inside a plane, there is no influence of the third dimension. Once again it is an oversimplification of a real sandy soil but it gives some hints on the micro-reorganizations of granular material because the 2D geometry allows to directly follow the motion of each grain by simply using a camera above the setup. More precisely the grains are not beads but small metallic hollow cylinders whose form is adapted to the 2D geometry: the axis of the cylinders are perpendicular to the plane of motion The cylinders have two different outer diameters d_1 = 4 mm and d_2 = 5 mm. Mixing 2 types of grains leads to a disordered granular media by avoiding crystallization, *i.e.* regular stacking of the grains. This is done in purpose for obtaining generic results. The inner diameters of the cylinders are also different, which allows a proper determination of the type of grains for further image analysis.

Around 4,000 such grains in an equal proportion in mass of the two types of cylinders are piled up onto an inclinable plane. All cylinders have a 3-mm height and lye on this plane (a low frictional glass plate allowing a backward illumination) without rocking. The lateral and bottom walls are made of Plexiglas and delimits a rectangular frame of L = 26.8 cm (54 d_2) width and an adjustable height of typically H = 34.4 cm (70 d_2). The 2D packing fraction c defined as the ratio of the surface of grains to the total surface they occupy is then c = 0.749 ±0.004.

For the experiment the bottom plane is tilted at an angle 0 such as to control the confinement pressure inside the granular material by an effective gravity field $g\sin 0$ where g is the gravity acceleration. A value of 0 = 33° is chosen for being larger than the static Coulomb angle of friction between the grains and the glass plate which is around e = 20° ($\mu_{grain/glass}$ = tan e is the static friction coefficient between grains and glass). Therefore the grains spontaneously move downward if they have the possibility to find a place below.

The intruder is a big grain of diameter d_2 located in the median part of the container at a 21.2 cm (*i.e.* 42 d_2) depth from the upper free surface. The intruder is attached to a rigid arm in Plexiglas (reinforced by metallic parts) moved by a translation stage and a stepping motor driven by a computer. The arm motion takes place along the median axis Y of the container and is parallel to the plane. In this report we use an intruder displacement value U_0 of a fraction of a grain diameter (U_0 = 1.25 mm H" d_2 /3) giving a typical strain less than 10^{-2}, far above the elastic limit but also far below the usual fully developed plasticity domain where shear bands appear.

The intruder is moving up then down to its first position and then again up and down and so on with always the same amplitude of displacement U_0, thus performing cycles of displacements in a quasi-static way. The up and down motions along the *Y*-axis are separated by rest periods during which pictures with a high resolution CCD camera (1280*1024 pixels^2) are taken. The image frame is centred slightly above the intruder and covers a zone of area 39 d_2*31 d_2.

In the following, we use the notation i for the index corresponding to the ith image just before the ith displacement of the intruder (upward or downward) and n for the cycle number with n = int[{i+1}/2] where int is the integer part. For each image i, the centre of each grain is determined with precision using the computation of the correlation on grey-levels between an image of the packing and two reference images corresponding to both grain types (d_1 and d_2). Note that the inner diameter of the cylindrical hole, which is different for each grain type (small or big), helps crucially for the proper determination of the centre and of the grain type.

Litter and Soil Organic Matter

There have been several syntheses of rates of accession and decomposition of litter in forests over the last two decades. Large amounts of litter fall from the above-ground parts of forests to the soil annually; this litter is a major determinant of nutrient cycles in forests. On a global scale, annual above-ground litter production in forests, woodland and wooded grassland is estimated at about 39 Pg of dry matter (1Pg= 10^{15}g), which is a little less than 50% of total global litter production. Annual above-ground litter accession is highest in tropical, broadleaved, evergreen forests at 15300kgha^{-1} and as low as 130 kg ha^{-1} in boreal, needle-leaved, evergreen forests.

On a global scale, average litter turnover time is about 5 years for litter and about 13 years for coarse woody debris (usually classified as materials >7cm diameter). As decomposition of forest litter is an important component of global C cycling, the potential impact of climate change on C release from litter has attracted considerable attention over the last decade.

In a recent review on a global scale of litter decomposition (measured as decomposition in the first year), Aerts (1997) concluded that climate (expressed as actual evapotranspiration) is a better predictor of the rate of litter decomposition than litter chemistry, although climate indirectly influences litter chemistry.

However, litter chemistry became the best predictor of decomposition rates within discrete climatic regions, especially in the tropics and in the Mediterranean region where the ratio of concentration of lignin to concentration of total N was the best indicator of litter decomposition. The strong correlation between decomposition rates and the lignin :N ratio of litter has been corroborated in a wide range of Canadian forests where 73% of the variance in decomposition was accounted for by a multiple regression of mean annual temperature, mean annual precipitation and the lignin: N ratio of the litter.

On a regional or global scale, the rate of litter decomposition in forests is predominantly controlled by climate (temperature and moisture, as shown in general terms by a strong negative correlation with latitude) and by the quality of litter in terms of how beneficial it is to the micro-bial community as a source of energy or nutrients. High concentrations of nutrients relative to stored energy promote more rapid decomposition rates, so that C: N and lignin: N concentration ratios are commonly used as variables for litter quality. Across the generally mesic forested biomes, climate and litter quality seem to be the best indicators of the rate of litter decomposition. However, the value of litter quality as a determinant of the rate of litter decomposition does not hold across all biomes, and is poorly correlated with the rate of decomposition in desert regions.

While accession of exogenous N to forested land has raised concern for increased decomposition of litter, reports of effects on decomposition rates vary. For example, Kuperman (1999) reported increased rates of litter decomposition (and of N mineralization) in white oak stands of the lower midwestern USA following increased N deposition. However, in a study encompassing a wide range of sites, Prescott (1995) found that increased N availability following fertilizer addition or deposition did not alter rates of litter decomposition in forests.

It seems clear that while exogenous N may increase decomposition in some forests, especially over the first year following litterfall, it may also increase the lignin content of litter and reduce long-term decomposition rates in other forests.

The reduced decomposition of N-rich litter may be due to the chemical formation of stable nitrogenous compounds from lignin by-products. These N compounds are thought to form by chemical condensation reactions and may be highly resistant to biological degradation. Current evidence therefore suggests that an increased rate of N deposition decreases the rate of C release in the latter stages of decomposition and stimulates C storage, both in increased woody biomass and through soil humus formation and, thereby, sequestration of C as soil organic matter.

Below-ground inputs of C and nutrients in forests may become the dominant pathway of nutrient return, especially in cold-temperate climates. Berg *et al.* (1998) studied root litter decomposition in coniferous forests from northern Scandinavia to north-eastern Germany and found little correlation between the loss of mass in the first year and climatic variables when different forest types were considered together. When the *Pinus* and *Picea* stands were treated separately, stronger correlations emerged with initial concentration of P and July temperatures explaining 71% of the variation in the rate of decomposition of root litter of *Picea* stands along the transect. Soil-warming experiments in a northern hardwood forest, where the forest floor was heated with a cable, showed that the CO_2 flux from leaf litter increased exponentially with increasing soil temperature.

These increased rates of CO_2 flux from litter decomposition may represent a short-term depletion of labile C with increasing temperature, so that the response diminishes over time. However, if global warming causes increases in the rate of biomass production and in the rate of litterfall, the rate of litter decomposition will increase and increases in the rate of CO_2 emissions from forest litter will be sustained.

Timber Harvesting, Nutrient Removal and Sustained Productivity

There have been many studies of the mass of nutrients in the above-ground parts of a wide range of forests, so nutrient removal in harvested timber can be assessed with a degree of certainty. For a range of forests of the eastern USA, Federer *et al.* (1989) calculated that a whole-tree harvest of a mature forest removes about 250kgha^{-1} of N, 20kgha^{-1} of P, 350kgha^{-1} of Ca (up to 600kgha^{-1} for oak-hickory forests), 40kgha^{-1} of Mg and 150 kgha^{-1} of K. Losses of P, Mg and K in nutrient removal and harvest-induced leaching for a single harvest were less than 3% of the total pools, and the loss of N and Ca was less than 8%, except for oak-hickory forests where the loss of Ca was 13-19%. Data for eucalypt forests in Australia are similar; losses of N and P due to timber harvesting ranging from 2% when only the stems are harvested up to 5% for whole-tree harvesting.

The magnitude of nutrient loss varies with nutrient, species and harvesting practice, including rotation length, intensity of harvest (debarking or not, stems or whole trees removed) and site preparation for regeneration (particularly whether the logging slash is burned). Judd (1996) simulated nutrient loss due to timber harvesting for a range of harvesting regimes and for two subgenera of the genus *Eucalyptus: Monocalyptus,* which includes the fibrous-barked stringy-barks; and *Symphyomyrtus,* which includes the smooth-barked gums. Judd's model excludes leaching losses. For natural forests with an 80-year rotation, Mg and K are always in positive balance and P always negative. N becomes negative when fire is used to prepare the site for regeneration. Losses of all nutrients increase rapidly as the rotation time decreases. In the extreme comparison, the Ca balance changes over 80 years from +557kgha^{-1} (fibrous-barked natural forest, 80-year rotation, logs debarked on site) to -2168 kgha^{-1} (smooth-barked plantation, 10-year rotation, logs not debarked), a fivefold increase.

This emphasis on the loss of Ca as a potential problem for sustained productivity was also raised by Turner and Lambert (1996) for plantations of *Eucalyptus grandis,* a smooth-barked eucalypt in the subgenus

Symphyomyrtus. About 60% of the total pool of Ca in a 27-year-old stand was held above ground.

Hopmans *et al.* (1993) concluded that Ca and B were the nutrients at greatest risk in managed, natural eucalypt forests in south-eastern Australia. In the study of forests of the eastern USA previously cited, Federer *et al.* (1989) calculated that the present rate of leaching will remove all the Ca in less than 1000 years. If losses due to short-rotation, whole-tree harvesting are added to these leaching losses, 50% of the Ca will be removed in only 120 years.

Clearly this rate of leaching of Ca in the forests of the eastern USA must be recent, otherwise the ecosystem would be entirely depleted. Federer *et al.* (1989) suggest that it has been caused by acid rain. The H^+ in solution is replaced by the metallic cations Ca^{2+}, Mg^{2+}, K^+ and Na^+. If the input of H^+ in acid rain as H_2SO_4 and HNO_3 was one-tenth the current rate, the metallic cations would be in long-term balance, and this is likely to have been the pre-industrial condition.

By comparison with losses of Ca due to timber harvesting, the losses of N and P (the elements previously and generally considered to be the more limiting and therefore of most likely concern) are relatively small. To put these losses of N and P into a different perspective, we have calculated loss in terms of a change in concentration in the surface 30 cm of soil. For N, the change in concentration of N in the soil over 100 years is a mere 1–2ppm annually (compared with total N about 5000 ppm) and is an order of magnitude less for P, 0.05-0.15 ppm annually (total P about 300 ppm).

In our example of harvesting a eucalypt forest of high productivity on a rotation of 100 years, N balance is restored if the rate of symbiotic N_2 fixation is 3.6-6.8 $kgha^{-1}year^{-1}$, depending on the intensity of harvesting. *Acacia* species are a natural component of both under-storey and shrub layers in eucalypt forests, and where they grow densely rates of N_2 fixation of 20-30 kg $ha^{-1}year^{-1}$ have been recorded. For most forests, however, the distribution of N_2-fixing species is irregular, and that makes the estimation of an areal rate of N_2 fixation very difficult. If we are to replace N losses by managing N_2-fixing species, we need much more research on managing the rate of N_2 fixation by these species relative to the long-term economic value of their N input.

Replacing the loss of P: weathering and pools of P

Resolution of a decrease in P of 0.05-0.15 ppm annually due to harvesting our eucalypt forest on a 100-year rotation is a futile quantitative exercise. Apart from analytical uncertainties in assessing the pool of plant-available P and the rate at which it is sustained over 100 years, there are basic questions that remain unanswered.

1 The rate of rock weathering is a 'ball-park' estimate; how can it be improved?

2 Does P move from lower-class (or less available) reserves, in both surface soil and subsoil, to higher class at the rate of 0.05-0.15 ppm annually?

3 Are tree roots chemically active in obtaining P from lower class reserves? For example, the roots of some forest tree species exude low-molecular-weight organic acids that may play a role in solu-bilizing P.

Replacing Nutrient Losses in Timber Harvesting with Fertilizers

The correction of a nutrient deficiency by the addition of an appropriate fertilizer is a simple and relatively economic business. However, compared with agriculture the use of fertilizers in forestry is relatively recent. White and Leaf's (1956) comprehensive bibliography of fertilizer research in forestry contained a mere 700 entries. About 25% of the papers were concerned with nurseries and nursery practice, and some 10% gave attention to 'degraded' forests (also described as 'stunted', 'degenerated', impoverished', etc.) in central Europe (particularly Germany and the former Czechoslovakia) and to the reclamation or 'improvement' of 'useless' land (swamps, bogs, peats, heathlands, moorlands, *Calluna* soils) in the UK, Sweden, Norway, Denmark and Finland and the subsequent establishment of tree plantations.

The 1960s saw the start of extensive use of fertl-izers in forests. Tamm's (1968) historical view of the development of the use of fertilizers in Europe is worth recalling.

- 1865-1900: the recognition, starting with Ebermayer (1876), of nutrient cycling and nutrient demand in forests; almost no experimentation.
- 1900-25: trial-and-error field trials, with little analytical basis.
- 1925-60: scientific foundation for tree nutrition established; many experiments but limited practical application.
- 1960 onward: the start of large-scale application of fertilizers in some forests.

And, indeed, this was a large-scale expansion. For example, the area of fertilized coniferous forest in the north-western and south-eastern USA increased from zero in 1967 to 850 000ha in 1978. Baule (1973) estimated that the worldwide area of forest to which fertilizer had been applied doubled from 2 million ha in 1970 to 4 million ha in 1973. While these areas are impressive, we should remember that they are less than 0.1% of the world's forested land. In short, fertilizers are not used on most of the world's forests. The traditional view of tree nutrition in silviculture of natural forests has always been that 'the loss of nutrient elements through the exploitation of timber is not very serious if the leaves and twigs are left behind in the forest' and that 'while mineral nutrients are as indispensable for trees as for agricultural crops, the cycle [of nutrients] rarely breaks down under rational forest management, and so mineral elements "take care of themselves"'.

Evidence for a Decline in Productivity due to Nutrient Removal

The assertion that the loss of nutrients associated with timber harvesting in natural forests causes a *loss* of productivity is regularly made in the ongoing debate over forests and logging. However, we have not found any definitive quantitative evidence in the literature in support of this assertion, nor are fertilizers used regularly in most of the world's forests used for timber production.

In contrast to the natural forest, fertilizers are used regularly and increasingly in plantation forests around the world. We do not attempt here to cover the many and varied fertilizer prescriptions that have been developed but point to some relatively recent reviews of the use of fertilizer in plantation forests, including Miller (1990) for European forests, Allen *et al.* (1990) for southern pine forests of the USA, Boomsma and Hunter (1990) for *Pinus radiata* plantations in Australia and New Zealand, and Attiwill and Adams (1996) for eucalypt plantations in Australia, New Zealand, Argentina, Chile, Brazil, India, China, Portugal and South Africa.

Plantation forestry very often involves species planted well outside their natural range (*e.g. Pinus* in Australia and New Zealand; *Eucalyptus* in Brazil, South Africa, India, Spain and Portugal), often on soils that are viewed as marginal for most other uses.

Nutrient deficiencies and imbalances are usually obvious and can be readily diagnosed and remedied. Furthermore, plantation forestry is more easily managed as a farming process than is the management of a natural forest. In plantation forestry, we manage on rotations of 10 years or so for fibre and on rotations of 15-50 years or so for timber. In natural forests we deal with rotations of 80-150 years.

We establish and manage plantations for a specific purpose, whereas we manage a natural forest for all its benefits, such as water production, recreation, aesthetic well-being, conservation of fauna and flora, and timber. The costs of establishing and maintaining plantations are more readily related to a commercial benefit than are the costs of maintaining a natural forest.

Maintaining the productivity of natural forest requires very different scales of perspective, both in time and space, from those that apply in plantation forestry which, in turn, are very different from those that apply to agriculture. In agriculture, we can compare this year's crop with last year's, or with mean yields over the past 10, 20 or 50 years. In plantation forestry, we can compare the productivity of a given compartment with the productivity of the previous rotation. For most natural forests, however, we do not have the records that allow a comparison of the productivity of today's forests with those of 100 years ago. An often-quoted example of nutrient depletion is the 'second-rotation decline' of *Pinus radiata* plantations in south-eastern Australia. Plantations of pines were established on deep, white, podsolized dune sands that previously supported natural woodlands of no commercial value for timber production. After the first rotation, the litter layer that had accumulated under the pines was heaped and burned along with the logging slash. A serious loss of production between the first and second rotations was clearly recognized by detailed assessment as early as 9.5 years into the second rotation. Remedies included both the replacement of nutrients by the addition of fertilizers and the conservation of nutrients by the conservation of organic matter between rotations.

We argue strongly against the generalization of productivity decline due to timber harvesting based on the experience in the pine plantations of South Australia, a conclusion also reached in the recent review by Evans (1999), and we argue in particular against the extrapolation of this experience to the natural forest.

There are many considerations for the natural forest, including rotation length, intensity of harvest, the appropriateness of harvesting methods and silvicultural treatments relative to the ecology of natural regeneration, and the influence of other factors (*e.g.* Ca depletion due to increasing acid rain in some forests of eastern USA; Federer *et al.* 1989).

Nutrient Demand in the Natural Forest

The total cycle of a nutrient in a forest is the sum of all the fluxes of the nutrient between plant and soil *plus* that amount of nutrient needed for new growth. The total cycle is calculated as the sum of the amounts of nutrient: (i) cycled from plant to soil in litterfall; (ii) recycled within the plant; (iii) in inputs less that in outputs; and (iv) taken up from soil reserves. For a mature forest, the percentage contribution of litterfall to the total annual cycle decreases in the order Ca > Mg > N > K > P, canopy leaching K > Ca = Mg > N > P, internal redistribution *P* > N > K > Mg > Ca, inputs-outputs N = Ca > Mg > K = P, and net supply from soil reserves K > *P* > N > Mg > Ca. For all the nutrients discussed here, annual uptake by the mature forest from soil reserves accounts for less than 20% of the total annual demand.

Of course, the relative contributions of nutrient cycling pathways to the total cycle must change with age of the forest. Attiwill (1980) defined three stages of forest growth and nutrient cycling, and Miller (1981) related the response of forests to fertilizers within essentially the same three stages.

1. Up to 20 years or so, the forest builds its productive system; the major proportion of NPP is used in building the photosynthetic crown and the systems for transport of water, nutrients and assimilates. This period is dominated by uptake from soil reserves, and Miller

writes that this is the stage 'when responses to a wide range of fertilizer [additions] can be expected'.

2. The rate of increase of the productive system is slowed. An increasing proportion of NPP is stored as physiologically dead heartwood, internal recycling accounts for an increasing proportion of the total cycle of the physiologically mobile nutrients, and the dependence on uptake from soil reserves decreases. A response to fertilizer might only be expected if the canopy is thinned.
3. The trees reach maturity and the rate of biomass increase approaches zero. The major proportion of NPP is now shed as litter. Nutrients (particularly N and P) may be immobilized in the litter; additions of N and P may speed up decomposition but will have little effect on growth of the trees.

Does the addition of fertilizer increase the carrying capacity of a forest? For agricultural crops, forestry where productivity has been restored on degraded land, and plantations that have been established on previously unproductive (in terms of timber production) land, limiting factors are recognizable and some of them can be easily remedied.

For example, on some of the sandy soils in Australia, the success of agricultural and tree crops depends on additions of superphosphate and trace elements such as Cu and Zn. However, the productivity of ecosystems (including agricultural and forest crops) is up to an order of magnitude less than so-called 'potential' production. Remedies such as draining, ripping the subsoil, mounding, adding fertilizers and irrigating overcome some, but by no means all, the limitations.

However, for the natural forest and many plantations, it seems that an increase in carrying capacity due solely to the addition of fertilizers is unlikely. Nor is it likely that fertilizers will be used extensively to increase the rate of production. The efficiency with which trees use nutrients added as fertilizer is low, and growth responses are short-lived. It is more likely that fertilizers will be used at relatively low rates to maintain productivity and essential ecosystem processes such as mineralization.

Timber Harvesting, Soil Compaction and Sustainability

Timber harvesting and logging equipment compact the soil and therefore change the physical properties of soil. Compaction is due to a decrease in macropores to a depth of 30 cm or so, resulting in increases in bulk density, water-filled porosity, available water and soil strength, and in decreases in sorptivity, aeration porosity and hydraulic conductivity. Despite the magnitude of some of these changes (*e.g.* an increase in bulk density in the surface 0-10 cm of soil of 20-40%; Rab 1996), the effects of compaction on plant growth are varied. Where disturbance by machinery is greatest (*e.g.* skid trails and log landings) the effects of compaction are compounded by the removal of organic layers and topsoil and by the exposure of subsoil. Among tropical forests, for example, height growth of mahogany was unaffected by soil compaction, whereas growth of a number of species in Sabah, Malaysia, and in Amazonian Ecuador was reduced by compaction but responded more to the addition of fertilizers than to a range of treatments designed to ameliorate the effects of compaction. Soil compaction may affect root growth more than shoot growth.

Given that the snig tracks and log landings are properly revegetated, the physical properties of soils recover to precompaction conditions within 5-40 years, depending on soil type, organic content and intensity of compaction. Thus one could conclude that where recovery time is less than one rotation, the effects of compaction *per se* will not be cumulative.

However, the problem of loss of topsoil through bad forestry practice seems to be of greater and longer-term significance than the problem of compaction alone. Restoring the fertility and physical structure of topsoil will take centuries rather than decades.

As well as direct effects of compaction and removal or erosion of topsoil on the subsequent growth of trees, there may be direct effects on diversity. The topsoil is a store of seed of some plant species and of vegetative propagules for others. Significant decreases in number or abundance of plant species have been observed after logging in aspen forests in the northern Lake States region, USA, in rain forest in Sabah, Malaysia, and in mountain ash forest in Victoria, Australia.

In summary, conservative management of the soil to minimize erosion and compaction is fundamental to sound land management; it is as essential in forestry as it is in agriculture. However, the nature of soil compaction in forests differs from that in agriculture. The mass of the trees, the forces generated by wind and the increasing size of roots all lead to soil compaction. Equipment used in timber harvesting is heavy, and together with the movement of large logs exerts great pressure on soils. Finally, equipment used in timber harvesting does not traverse the land as uniformly as that used in agriculture, so that heterogeneity of compaction and soil disturbance is greater in forestry than in agriculture.

Furthermore, timber harvesting generally requires the construction of roads and logging tracks to reach new logging areas. These new roads and tracks are particularly prone to erosion. Siltation of streams due to roading plus logging may be orders of magnitude greater than siltation due to logging alone. Conservative management of soil in forest operations must therefore be all-embracing. Roads, skid trails and log landings must be carefully planned: steep slopes must be avoided, river crossings minimized, cross-drains created to disperse water, buffer zones established and maintained, and revegetation hastened. All these requirements can be met through strict adherence to, and monitoring of, codes of forest practice.

Assessment of the physical condition of a coupe after logging, especially the amount of soil disturbance, is targeted through sustainability indicators developed by the Working Group (1995) for the Montreal Process. Where fire is an essential component of forest ecosystems, these sustainability indicators must be sensibly interpreted. For example, *Eucalyptus regnans* (mountain ash) in Victoria, Australia, regenerates in nature only after bushfire, and the ash-bed effect (the burning of all shrub, ground and litter layers and heating of surface soil) is essential to germination, survival and growth. In commercial forestry operations, the trees are clearfelled and the logging slash is burned to prepare the seed bed. Subsequent seedling growth is always significantly greater on burned soils than on unburned soils, including those where a seed bed has been mechanically prepared. Sustainability indicators that emphasize retention of litter layers to protect topsoil will therefore have little basis in ecological processes when fire is involved.

SOIL ACIDITY AND PLANT GROWTH

As soil becomes more acid the availability of aluminium to the plant increases to toxic levels, while the availability molybdenum, phos-phorus, magnesium and calcium decreases. In weakly weathered soils manganese toxicity will develop as the soil becomes more acid.

These changes in availability of aluminium and plant nutrients cause most of the effects on plant growth attributed to acidic soils. Soil acidity may also reduce nodulation and nitrogen fixation by some legumes, decrease rates of mineralisation of organic matter and de-crease microbial activity in general.

Aluminium (Al)

Available aluminium increases as pH falls. The available Al at a given pH varies between soils with more aluminium available in well weathered soil than in a weakly weathered soil.

The principal effect of aluminium toxicity is to restrict root development by disrupting the structure and function of roots. Symptoms of aluminium toxicity in the leaves are similar to those of phosphorus deficiency and moisture stress as it decreasing the ability of plant roots to grow into acidic soil and extract nutrients and moisture.

Crops wilt more quickly during dry periods and an overall reduction in water use efficiency of both pastures and crops occurs. Plants vary in their tolerance of aluminium. Tolerant plants can survive high levels of aluminium in the soil. These plants are also affected as high levels of aluminium immobilise phosphorus in the soil and can affect the uptake and utilisation of calcium and magnesium. Plants vary in their tolerance to soluble aluminium. The tolerance of a number of commonly grown agricultural crops and pastures grown in Australia. The plants are arranged into four tolerance classes.

Manganese (Mn)

Both toxicities and deficiencies of manganese can occur in plants. Soils most likely to develop manganese toxicity are weakly weathered soils that are high in easily reducible manganese oxides, such as red earths and red brown earths and more highly weathered soils that are high in iron and aluminium oxides that adsorb and accumulate Mn^{2+} . Not all plants are sensi-tive to Mn toxicity. In those plants that are sensitive, Mn toxicity causes chlorosis and crinkling of the leaf margins. Deficiencies are only expected where there is an absolute deficiency as in the case of some siliceous sands, or in soils that are very high in calcium carbonate (*e.g.* > 50 per cent).

High Fe soils do tend to accumulate Mn rather than being depleted in Mn during soil development. This could be a problem in Brazil but the limited literature seen by the authors suggest that an Mn deficiency may be a bigger problem. Increasing the soil pH to above 5.5 with lime can lower high levels of available manganese in the soil. On the other hand Mn deficiency can be accentuated by liming.

Molybdenum (Mo)

Molybdenum is a trace element required for nitrate conver-sion to the ammonium form used in plant proteins and in the nitro-gen fixation process. Where the pH of a soil is below 5.5 in soils naturally deficient in Mo an application of 50–100g Mo per hectare is usually sufficient to eliminate Mo deficiency for 5 to 10 years. In soils high in iron and aluminium oxides more may be required.

Calcium (Ca)

In the experience of the authors most agricultural soils contain an adequate supply of calcium as a nutrient for the crops and pastures grown. Where there is a moderate calcium deficiency, it is seen first in parts of the plant that are furthest from the main flow of water within the plant such as production of peanuts. Low levels of soil calcium can also adversely affect the nodulation of sub-terranean clover.

Short term severe deficiencies can cause petiole collapse of young expanding leaves. Examples of very severe calcium deficiency are most likely in soils with a pH less than 4 that are sandy and low in organic matter, or where there has been excessive use of highly acidifying ferti-lisers. Under these circumstances the exchangeable calcium can drop to concentrations less than 40 per cent of the exchangeable cations other than aluminium and this can only occur in soils with a low cation exchange capacity.

As these soils usually have very high levels of soluble aluminium, all but the most acid tolerant plants, such as sugar cane, are killed by the aluminium before the symptoms of calcium deficiency become apparent. The symptoms are stubby, weakly branched and discoloured roots and dead shoot growing points. The root symptoms are difficult to distinguish from symptoms of aluminium toxicity.

Magnesium (Mg)

Losses of production in crops and livestock due to a nutri-tional deficiency of magnesium are most unusual. In Australia most of soils have adequate supply of Mg in the subsurface soil layers if not in the surface layers. Excess Mg can in some circumstances cause loss of soil structure.

Phosphorus (P)

Phosphorus is absorbed onto iron and aluminium oxides where it is less available to the plant and this effect increases as soils become more acid. Thus acidic soils high in iron and aluminium oxides need to contain higher concentrations of phosphorus to avoid P deficiencies in plants.

The Process of Acidification

The source of the acids in the ecosystem is principally the carbon and nitrogen cycles within the eco-system.

The Carbon Cycle

Description of the C cycle starts in the plant leaves with the creation of sugars from carbon dioxide (CO_2) and water using the energy of the sun through photosynthesis. These sugars are con-verted in part to organic acids that dissociate into H^+ ions and organic anions. The negative charge on the anions is balanced by the "base" cations, Ca^{2+}, Mg^{2+}, K^+ and Na^+ to form salts of organic acids.

These base cations are taken up through the root. The H+ ions are consumed in metabolic reactions or are excreted at the root surface in exchange for excess cation uptake.

Note that the plant requires that the charge remain constant across the root mem-branes when taking up water and nutrients but does not require that a constant acid-base balance be maintained. The result is that the plant material becomes alkaline and the soil becomes more acid.

The internal pH balance within the plant is maintained through the Malate – Pyruvate buffer within the plant. When the plant loses its leaves, or dies, the plant tissue be-comes part of soil organic matter that in time breaks down to become CO_2 and water again, and releasing the alkalinity stored as organic anions. Fire is the same process but occurs far more quickly. The alkali released in the break down of the organic matter will neutralise the acid in the soil.

As the alkali is accumulated in the above ground biomass an equal amount of acid is added to the soil. The return of biomass to the soil neutralises the acid as it breaks down. Some acid will be utilised in the weathering process otherwise the eco-system will retain an alkali/acid balance that, if the system was totally closed, would continue indefinitely. However if plant material is removed from the eco- system as in the harvest of grain or the making of hay then the eco-system is left more acid.

The Nitrogen Cycle

The description of the N cycle starts with nitrogen being taken from the air by legume plant/Rhizobia symbiosis and incorporated into the plant. When the plant dies the nitrogen becomes part of the organic matter that in turn breaks down releasing ammonium nitro-gen.

The ammonium nitrogen through mineralisation and nitrifica-tion becomes nitrate nitrogen with a net release of acid into the soil. As the nitrate nitrogen is taken up by the plant root it excretes an equal amount of anion that will neutralise the acidity as the charge across the root membrane must remain constant. If all the nitrate nitrogen produced is taken up by the plant then the acid alkali balance in the eco-system will continue inde-finitely. However if any of the nitrate is leached from the root zone before the plant can take it up then the soil is left more acid. Some nitrogen may return to the atmosphere through a further step called denitrification.

This is an alkaline process that balances the acid produced during the mineralisation and nitrification process. Note that contrary to some older texts, the removal of the "base" cations, Ca^{2+}, Mg^{2+}, K^+ and Na^+ is not the cause of acidity in the soil but rather the result of acidification of the soil by acids added in the C and N cycles within the eco-system.

Creation and Destruction of the Stable State' of Soil Acidity

In nature most eco-systems have evolved to an almost stable state where the acidity/alkalinity balance changes slowly towards becoming more acid. In this natural stable state the eco-system is producing a small amount of acid that releases sufficient cations from the parent material to maintain fertile soil. For this publication the term "stable state pH" will refer to the alkali/acid balance in an eco-system that has evolved to the point where changes are very slow.

The natural processes that control acid addition in the natural eco-system:

- The age of the soil.
- Climate, particularly rainfall, affects the intensity of leaching and of erosion and hence the opportunity for loss of alkaline materials.

- The vegetation affects acidification through variations in N fixation capacity, in the ability to absorb and recycle nu-trients from and in the soil profile, in its ability to minimise the leaching of nutrients through effects on water use and nutrient 'stripping' from the leached water. In addition diffe-rent plants contain different concentrations of anions and this affects the amount of alkali lost when plant material is removed and alkali returned to the surface from the deeper soil layers.
- Soil parent materials affect the rate of acidification due to differences in their acid neutralising capability or buffering capacity. Basalt, and often alluvium, have a high neutra-lising capability (also called a pH buffering capacity) com-pared to previously weathered materials such as a sandstone which may contain 98 per cent quartz (SiO_2).

In Wagga Wagga the stable state pH(0 to 10 cm) is between 5 and 6, increasing with depth to about 8 at 1 metre for soils that have not been disturbed by farming. These soils are generally duplex soil and support open woodland vegetation. This stable state pH has developed over more than maybe 30,000 years and has as part of its genesis the regular burning by the aboriginals who inhabited this land for more than 40,000 years. When the land is turned over to agricultural production acidification rates increase and the 'natural' stable state pH is destroyed.

The normal agricultural practices of fertilising and harvesting along with the increase in leaching of nitrate nitrogen accelerate the acidification of agricultural lands. In addition a build up in organic matter takes organic anions out of the eco-system leaving it a little more acid but this is reversed if the organic matter is reduced later. An increase in erosion of the fertile topsoil will remove this organic matter leaving the soil more acid.

The challenge to farmers and their advisers is to establish a stable state pH for farming if it is to continue for 100's and maybe 1,000's of years. This will mean managing the processes that acidify the soil such as fertilizers, product removal and increased nitrate nitrogen leakage away from the root zone and applying an alkaline material, most likely applying finely ground lime.

In Wagga Wagga the stable state pH that cropping farmers strive to achieve is pH between 5.0 and 5.5 by applying finely ground limestone. Most plants and soil biota are not affected by the acidity at this soil pH and the net movement of $H+/OH-$ down the profile is in balance. This stable state pH is economically sustainable as their cropping/pasture rotation is based on plants that are sen-sitive to acidity and the purchase of lime comes from, and maintains, the farm profits. In an experiment at Wagga Wagga the long term effect of tillage, crop rotations, lime and fertilizer on soils has been running since 1979. This experiment indicates dramatic effects on pH (0-10cm) occurred in the first ten years but this has now stabilised. Acid addition is now thought to be equal to the ongoing weathering rate seen as a 7 per cent reduction in fine clay fraction over 20 years. To the east of Wagga Wagga in the non-arable grazing lands lime application does not increase profits sufficiently to cover the cost.

The farmers who are establishing a new stable state pH grow perennial, acid tolerant pastures and their agricultural practices are less acidifying. These farmers have a stable state pH that is less than 5.0 and often less than 4.8. While acidification rates are lower than at 5.5 it is not sustainable as the whole profile is acidifying.

Distribution of Soil Acidity Over the Soil Profile (Ph Profile)

To understand how to achieve a fertile stable state pH profile it is necessary to understand the factors that control the distribution of soil acidity down the profile. Planning a cropping or pasture pro-gram that will establish a sustainable, fertile pH profile requires management of the acid being added to each layer within the profile.

Examples of ways that pH profile influences crops and pastures in an acidic soil:

- Acid sensitive plants will be affected if the pH is below 4.7 beyond 20 cm. The root of an acid sensitive plant can often break through 10 cm of inhospitable acidic soil is less likely to penetrate 20 cm of acidic soil.
- The effect of lime moves down the profile very slowly and only when the pH in the layer above is greater than 5.5.
- The response to liming is affected by the pH profile below the limed layers.
- Where sub-surface acidity occurs it may take a number of years for the economic response to liming to be realised so the pH profile is important in planning a crop and pasture rotations in association with a liming programme.

Development of a Ph Profile

Different soil profiles develop under different agricultural, forestry and natural eco-systems. For example in many systems the surface 5 cm has higher pH than the sub-surface layers (5-20 cm). In undisturbed soils around Wagga Wagga, for example, the top 5 cm has a pH of 5.5 to 6, a pH of 4.5 to 5.5 in the 5 to 20 cm layer with the pH rising again in lower root zone due to net alkali addition. In other systems the reverse may occur. Where net acid addition occurs on a whole profile basis, the long-term effect is usually a strongly acid profile to depth, with a slightly less acid layer in the top 2-3 cm.

Subsequently leaching, diffusion and soil movement or mixing, can redistribute hydrogen and hydroxyl ions and their che-mical equivalents (aluminium, carbonate and bicarbonate ions) within the profile. If it is difficult to avoid the development of a highly acid pH profile to depth over laying a less acid deep sub-soil. In the natural eco-system this happens very slowly but where acidification is accelerated the profiles will become acid to depth quickly unless finely ground lime is incorporated to at least 10 cm to neutralise any acid addition.

The factors that affect the development of the soil pH profile are:

- Leaching of nutrients affect the location of acid and alkaline reactions in the N and C cycles, and the pattern of excre-tion of H^+ and OH^- (or HCO^{3-} and organic anions) by roots in response to imbalances in cation and anion uptake.
- Net acid addition can occur close to the top of the profile if nitrate formed from nitrification is leached. Acidity and alka-linity can be redistributed within the soil profile by mass flow in water, by diffusion or by soil movement. In general the movement by water will be acid if the pH is below 5.0 and net alkaline if the pH is above 5.5.
- The alkaline effect of applied lime moves slowly down the profile (pH > 5.5), unless the surface pH is high (pH 7) and biological activity is high.
- Plant transport of nutrients and organic anions onto the soil surface.
- Some worms and ants will bring soil up from depth and will transport lime down the profile both by active movement and faster leaching down the holes they create.

The locations in the soil profile where N and C cycle acids and alkalis are added. The primary processes of acid addition to the soil from the N and C cycles are:

- Nitrification.
- Dissociation of organic acids in the soil (from organic acids produced in the soil by microflora or organic acids added to the soil in plant residues).
- Immobilisation or volatilisation of ammonium ions in the soil.
- Excretion of hydrogen ions by roots in response to uptake of more cations than anions.

The primary processes of alkali addition to soils (or hydrogen ion consumption) are:

- Denitrification, nitrate immobilisation by microflora.
- Mineralisation of organic N to ammonium.
- Adsorption of hydrogen ions by organic anions in organic matter added to soils with lower pH than the organic matter.
- Oxidation of organic anions to carbon dioxide and water.
- Excretion of OH or HCO_3 by roots in response to greater uptake of anions than cations.

Therefore the net distribution of acids and alkali addition to the soil profile depends on the distribution of all the above primary processes of addition of acids and alkalis. Among the quantitatively dominant processes, the oxidation of organic anions (alkaline) and ammonification (alkaline) usually occur near the soil surface where the organic matter is added.

The location of the dominant acid process, nitrification, depends on the depth to which ammonium is leached before nitrification occurs. Finally another quantitatively important process, the pattern of ex-cretion of acid or alkali by roots, is affected by the relative availability of the nutrient cations and anions in a given soil layer.

The most useful simplification of this balance is that roots usually absorb more non-N cations than non-N anions so can be expected to excrete some acid where N availability is low in a soil layer. Where N availability is high, rates of acid excretion can be expected where most of the soil N is in the ammonium form. Conversely moderate to high rates of alkali excretion can be expected where most of the soil N is in the nitrate form.

Therefore ecological processes such as the leaching of ammonium after ammonification and the leach of nitrate before absorption by roots, affect the distribution of acid addition to the soil profile. The leaching of NO_3–N to depth within the profile before the plant takes it up is one example of how the leaching of nutrient can affect the pH profile. It will leave the point where the nitrification took place more acid and will increase the pH at the point where it is taken up. While this could be a method to correct the acidity in the sub soil it is not always certain that the NO_3–N will be taken up and not leached from the root zone.

Indirect Effects and Soil Acidification

Effects on epicuticular wax, litter decomposition in the upper soil horizon and cation leaching from foliage are all potentially damaging. However, for both sulphur and nitrogen pollution soil-mediated effects are probably of overriding importance. The importance of soil processes and, in particular, the concept of the 'mobile anion effect' and associated 'aluminium' toxicity were developed by Ulrich and his cowork-ers in the early 1980s.

It is hard to underestimate the importance of this work: it formed the conceptual basis for soil critical load calculations now used to focus scientific understanding of pollutant impacts and to guide agreed international deposition and emission targets.

In brief, the mobile anion effect suggests that soil solution pH is usually buffered or controlled by the exchange of protons (H^+) with exchangeable base cations (Ca^{2+}, Mg^{2+}, Na^+ and K^+); these are normal cation exchange processes. If strong acids (H_2SO_4) enter the system, sulphate (SO_4^{2-}) and base cations will be present in soil solution and will be lost in soil drainage water. However, if the inputs of strong acid anions (sulphate and perhaps nitrates) persist, soil acidification continues until buffering of pH in soil solution is no longer via cation exchange mechanisms but rather by exchange with aluminium ions (Al^{3+}) in the $Al(OH)_3$ pool in soil solution (Al buffering range when $pH<4.2$). Soils then acidify (H^+ concentrations increase), Al concentrations in soil solution increase and percentage base saturation decreases with loss of Mg^{2+}, Ca^{2+} and Na^+ along with SO_4^{2-} in soil leachate.

In even more acid conditions there is a further Al/Fe buffer range ($pH< 3.8$). The resulting free Al^{3+} ions are toxic to roots and the loss of base cations and overall change is destabilizing to soil nutrient cycling. Full soil chemistry models are very complex, but an insight into the fundamental process is important since it informs understanding of soil status and of suitable management. There is now a very large body of data to support these ideas and they can be regarded as established science.

It is now clear that the atmospheric deposition of sulphur and nitrogen have a strong influence on the status of soils and on forest nutrition. Sulphur and nitrogen depositions and forest nutrition are strongly associated globally, although depositions and soil sensitivity to damage vary geographically. Recently, policy on emission abatements has been guided by the critical loads approach rather than, as previously, by an ill-defined concept of protecting ecosystems from damage.

For forest ecosystems, soil critical loads can be set using a mass balance equation such that deposition of strong acid anions does not exceed the rate of release of base cations by weathering of minerals (after allowing for cation loss from soil in nutritional uptake of cations and nitrogen) and an allowable leaching of alkalinity. In acidified soils low availability and uptake of base cations may limit tree growth, and the ratio between the divalent base cations (Ca and Mg) and Al in soil solution reaches a critical value (a molar ratio of 1.0).

The nutrient uptake is suboptimal, root damage can occur from free Al in soil solution, tree growth is reduced and visible forest damage may start to occur. The mass balance equation can be used to calculate S and N depositions (critical loads) below which (Ca+Mg)/Al is greater than 1.0 so that forest soils are protected from 'damage'. Information is now becoming available on S and N critical loads throughout Europe and initial work is being done on the soils of South-East Asia where major problems may occur over the next few years. The above summary is sufficient to indicate the strong link between acidic deposition, forest nutritional status and the ability of forest soils to sustain successive rotations without loss of their capacity to supply base cations and nitrogen. Evaluation of the status of forest soils and the consideration of suitable management options (*e.g.* liming) becomes possible when these mechanisms are understood.

Phosphorus

Phosphorus occurs in highly mobile forms within plants and, like N, is withdrawn from plant parts during senescence; internal redistribution of P accounts for 50% of the total annual above-ground cycle.

The half-life of P during decomposition is 4.1 years, *i.e.* P is immobilized throughout decomposition to a greater degree than N. A negligible amount of P comes in with the rain.

P as H_2PO_4- is adsorbed by clay colloids and is then highly immobile in the soils so that losses of P to streamwater are negligible. However, in forest soils more than 50% of the total P in surface soils may be in organic form. Mineralization of organic P is therefore fundamental to ecosystem function, and P availability is 'determined by competition between biological and geochemical sinks for phosphate anions'.

Whereas methods for estimating rates of N mineralization have been developed and are in general application, there are no methods for routinely estimating the rate of P mineralization.

The difficulty is that whenever H_2PO_4- comes into solution, there is a surface to adsorb it. This competitiveness between biological and chemical sinks for P remains a priority area of research in quantifying the long-term sus-tainability of P supply for forests.

Management for Maintenance of Soil Biology and Nutrition

The aim of management should be to create balanced organic matter and mineral budgets. It should ensure that, over several years (a complete crop rotation), soil organic matter is not depleted and that nutrients added equal or exceed those removed by cropping or lost in various ways.

When managing organic matter farmers should recognize that the effects of animal and human manure, sewage sludge and plant residues last longer than those of green manure crops. Green manures, though valuable, usually last only one or two seasons because they are incorporated before they are mature and lignified.

Juo and Payne (1993) advise: 'In spite of the many proven benefits of soil organic matter, its management and recycling in an intensified, modem agro-ecosystem must necessarily revolve around two fundamental characteristics of the system, namely, the availability of organic material at the farm level, and the economic incentive for conserving and recycling organic matter'. To this may be added consideration of the benefits and costs of treating and transporting human wastes from cities. On-site organic matter, organic material brought in from outside and industrially-produced fertilizers will be used according to their benefit cost ratios and the attitudes of governments and crop managers. Tillage reduces the frequency of VAM, at least in invertebrates. Tillage reduces earthworm populations 10-30-fold, both killing them directly and destroying their burrows. Haines and Uren (1990) show that differences in earthworm populations between tillage treatments and stubble burning are reflected by 50% differences in numbers of soil pores.

Tillage also effects the location, numbers and activity of microorganisms within the soil, depending on the type of tillage. For example a mouldboard plough that inverts topsoil has more effect than minimal tillage. Doran (1987) found microbial biomass and potentially-mineralizable nitrogen to be 54% and 37% higher respectively in the top 7.5 cm of no-till than ploughed soil.

These would be critical differences if translated to the semi-arid tropics. Microbial and fungal biomasses may be 20-70% higher after 1 and 2 years no-till. However, conventional tillage, by burying topsoil, may cause microbial populations and their activities to be higher than under no-till at depths of say 7.5 to 15 cm. Differences in soil organisms (*e.g.*, microbial biomass) under differing tillage treatments are seasonal, generally being greatest in the period when the weather is most favourable for soil organism activity and plant growth. Rasmussen and Collins (1991) summarize many data on the impact of tillage on soil properties and conclude that stubble-mulching with zero tillage conserves up to 2% more organic matter per year than ploughing.

Importance of Soil Biology

The activities of the wide range of organisms in soil play a pivotal role in both natural and managed ecosystems. Their processes of organic matter breakdown contribute to the soil's health – its stability, permeability, ability to retain nutrients and make them available for plant uptake.

Soil biological, physical and chemical processes are interrelated and all contribute to plant productivity. The level of soil biological activity is therefore affected by the soil type, but it also depends on the management practices used, particularly the management of organic matter, especially carbon. Changes that are made to the chemical and physical environment in soil will therefore influence the biological processes and subsequently the contribution they make to the soil's fertility overall.

Issues Associated with Soil Biology

Knowledge of soil biological processes can support decision making aimed at achieving sustainable use of agricultural land. Soil biology is a complex field, however, and research continues to uncover new facts concerning the organisms themselves, their processes and factors that affect them.

Additionally, management of soil biological processes is difficult to do precisely due to the differing parameters of each individual situation, such as soil type and land use. A certain land management practice may also affect one group of organisms, but not others. Consideration of the mass of microorganisms as a whole (microbial biomass) is therefore not sufficient for a complete interpretation of the effects of land management on soil biological fertility.

Measurement of soil health in terms of biological fertility is also a complex and at present, relatively expensive process. More attention has been usually given to the management of the soil chemical and physical environments. As a result, a host of inexpensive and simple tests are available to land managers. Incorporating soil biological processes into farming systems will require a more holistic and longer term approach to agricultural land management.

Benefits of Managing Soil Biology

Agricultural land management practices alter aspects of soil chemical and physical fertility with consequences for soil biological processes and vice versa. Both fauna and microorganisms contribute significantly to chemical transformations in the soil and influence their physical surroundings to various degrees.

Organisms on and around plant roots have major influences on plant nutrient availability and some form specific associations with legumes, which greatly influence the C:N ratio of plant residues in soil. Associations between agricultural plants and fungi known as arbuscular mycorrhizas have the potential to increase the efficiency of use of phosphorus in agricultural ecosystems as well as improve soil structure.

Current research includes understanding some of the undesirable members of the biological population, such as root pathogens. The role of mycorrhizal fungi in preventing and reducing the effects of salinity is also being investigated. Stubble management is another area of research.

Importance of Nitrogen in Plants

Of all the essential nutrients, nitrogen is required by plants in the largest quantity and is most frequently the limiting factor in crop productivity.

- In plant tissue, the nitrogen content ranges from 1 and 6 per cent.
- Proper management of nitrogen is important because it is often the most limiting nutrient in crop production and easily lost from the soil system.

Nitrogen Forms and Function

Forms of nitrogen available for plant uptake:

- Ammonium
- Nitrate

Functions of nitrogen in plants:

- Nitrogen is an essential element of all amino acids. Amino acids are the building blocks of proteins.
- Nitrogen is also a component of nucleic acids, which form the DNA of all living things and holds the genetic code.

- Nitrogen is a component of chlorophyll, which is the site of carbohydrate formation (photosynthesis). Chlorophyll is also the substance that gives plants their green colour.
  - Photosynthesis occurs at high rates when there is sufficient nitrogen.
  - A plant receiving sufficient nitrogen will typically exhibit vigourous plant growth. Leaves will also develop a dark green colour.

The Nitrogen Cycle

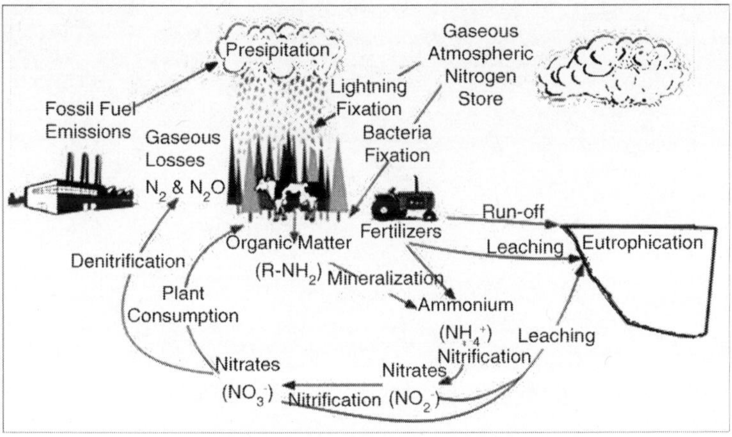

Fig. The Nitrogen Cycle

Gains of Nitrogen to the Soil:

- *Biological and Atmospheric Fixation*: Conversion of atmospheric nitrogen to ammonium which is subsequently available for plant uptake
- Direct additions of commercial and organic fertilizers

Transformations in the Soil:

- Mineralisation: Conversion of organic nitrogen to ammonium
- Nitrification: Conversion of ammonium to nitrate

Losses of Nitrogen from the Soil:

- *Denitrification*: Conversion of nitrate to atmospheric forms of nitrogen
- *Volatilisation*: Loss of gaseous ammonia to the atmosphere
- Run-off
- Leaching
- Consumption by plants and other organisms

Nitrogen is a very dynamic element. It not only exists on Earth in many forms, but also undergoes many transformations in and out of the soil. The sum of these transformations is known as the nitrogen cycle. Organic Nitrogen

Various compounds Organic nitrogen must be converted to ammonium before it is used by plants.

This conversion occurs with time and is known as mineralisation.

Though complex, the nitrogen cycle:

- Helps us to understand the complex relationships that exist between the many forms of nitrogen
- Provides us with insight pertaining to the availability of ammonium and nitrate, which are the only nitrogen forms usable by plants
- To understand the many ways in which N may be lost from the soil

In this section, we will discuss eight major transformations of nitrogen in the soil: nitrogen fixation, mineralisation, immobilisation, nitrification, denitrification, volatilisation, and leaching.

Nitrogen Fixation

Although atmospheric nitrogen gas(N_2) makes up approximately 78 per cent of the air, it cannot be directly used by plants.

Instead, atmospheric N_2 only becomes available to plants through three unique processes. The final product of each of these processes is ammonium, which is then available for plant uptake.

The three processes which convert atmospheric nitrogen to ammonium:

- Biological nitrogen fixation
- Chemical nitrogen fixation
- Atmospheric addition

Biological Nitrogen Fixation

Certain soil organisms have the special ability to convert atmospheric nitrogen to ammonium. These organisms include several species of bacteria, actinomycetes, and cyanobacteria.

In the soil, nitrogen fixating organisms can form special relationships with plants, called "symbiotic" associations. Symbiotic is a term that means "living together." Although a symbiotic relationship can be antagonistic, the symbiosis that occurs during biological nitrogen fixation is generally mutual and beneficial.

Legume-Rhizobium Symbiosis

The most abundant symbiotic relationship in nitrogen fixation forms between legumes (*i.e.,* alfalfa, soybeans, etc.) and the *Rhizobia* bacteria species.

- As the roots of legumes grow, *Rhizobium* bacteria infect the root hairs where they begin to multiply.
- As a response to this colonisation, the legume forms nodules, which are structures that form around the *Rhizobia*.
- Within these nodules, *Rhizobia* bacteria are able to continue multiplying and converting the N_2 from the soil air to ammonium.

- However, the presence of nodules is not a sufficient indicator that nitrogen is being converted to ammonium. Active and effective nodules are generally greater than 2 mm, have pink to red interiors, and concentrate around the tap root.
- On the other hand, non-effective nodules are generally smaller in size with white, green or brown interiors.

The image below shows cross sections of soybean (*Glycine max*) nodules. Nodules in the first row are highly effective at converting N_2 from the atmosphere into ammonium.

The nodules in the second row are moderately effective at biological nitrogen fixation, while the bottom row of nodules do not fix any nitrogen at all. Notice the lack of colour in the interior of these nodules, which indicates that they do not have an active 'nitrogenase system'. A 'nitrogenase system' is the bacterial enzyme that is necessary to convert N_2 gas into ammonium through this biological process.

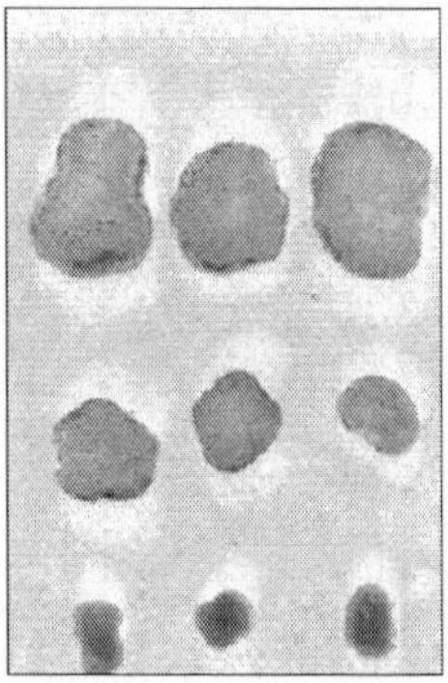

Fig. Relationship between Colour and Effectivity of Nodulation in Soybean.

The next image shows how a well nodulated soybean root system looks when it contains many highly effective *rhizobia* in the soil or is applied as an inoculant.

Fig. Depiction of Effective Nodulation of Soybean by *Rhizobia*.

Specificity

Some *Rhizobium* species are only capable of nodulating a particular legume species and cannot successfully nodulate other legumes.

- For example, the *Rhizobium* that nodulates alfalfa is a different species from the *Rhizobium* that nodulates soybean.

This phenomenon is known as *Rhizobium* specificity. However, not all *Rhizobium* are legume-specific. Thus, some may nodulate a number of different legumes.

The figure below gives examples of some common cross-inoculation groups that may assist you in selecting the proper rhizobial inoculant for a particular legume host. The proper combination of *rhizobia* and legume will result in the optimal nodulation and most nitrogen fixation.

From this figure, we see that using soybean *rhizobia* with a soybean plant forms an effective symbiosis, while using a soybean *rhizobia* with a *leucaena* plant does not. However, cowpea rhizobia is capable of nodulating both mungbean and peanut.

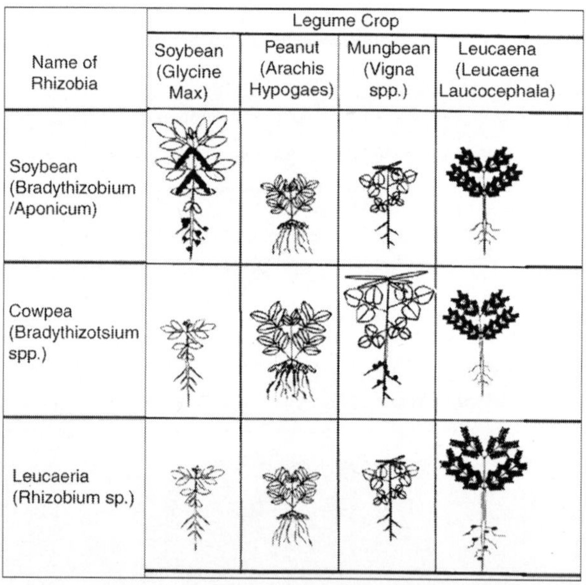

Fig. Specificity of Rhizobia for Successful Nodulation of Certain Legumes.

Biological Nitrogen Fixation Management Programme

The occurrence of the symbiotic relationship is heavily dependent upon a variety of soil conditions. If your programme incorporates nitrogen fixation, the following considerations can determine your success.

- First and foremost, the *Rhizobium* must be compatible with the legume. If your crop is *Rhizobium* specific, you must use the correct *Rhizobium* species.

- If your inoculum (which contains the *Rhizobium* bacteria) is applied to seeds, the procedures must be properly followed.
- Nitrogen fixation takes place when total soil nitrogen is insufficient. When sufficiently present, the plant will instead rely on the nitrogen available from the soil.
- *Rhizobia* are sensitive to any growth factor that limits root development. Such conditions as aluminum and manganese toxicities will limit inoculation.
- *Rhizobia* are influenced by mineral nutrient imbalances.
  - Low levels of calcium, phosphate, molybdenum under acidic conditions will limit nitrogen fixation.
  - Under alkaline conditions, phosphate, cobalt, boron, iron, and copper levels become a concern.
- Any growth factor (such as light, water temperature stresses or soil compaction) and any management factor (such as nutrient management, salinity) that detrimentally affects growth of the legume will detrimentally impact nitrogen fixation.

inoculation of Legumes with *Rhizobia* Increase Yield and Biological Nitrogen Fixation

This question is commonly posed by farmers who are deciding when to apply an inoculant to their crops. Since, inoculation is relatively inexpensive (less than $5.00/acre), growers should error on the side of caution and inoculate their legume crops unless they have good evidence that inoculation is not needed.The figure below is a conceptual model which integrates all the factors controlling biological nitrogen fixation. Additionally, it explains when the inoculation of legumes with *rhizobia* will result in an increase in plant growth and biological nitrogen fixation activity.

The model is based on the fact that if the plant's need for nitrogen is greater than the nitrogen that is supplied by both the existing soil nitrogen and the *rhizobia* already present in the soil, the inoculation of supplementary effective *rhizobia* will result in increased yield and biological nitrogen fixation.

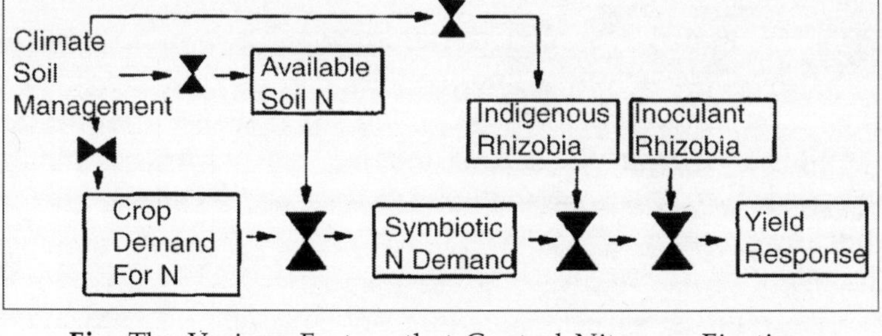

Fig. The Various Factors that Control Nitrogen Fixation.

Amount of Nitrogen Fixed by Legume

When nitrogen is converted to ammonium during biological nitrogen fixation, ammonium becomes available to the legume and the microorganism that fixes it.

Typically, the bacteria can fix anywhere between 20 and 80 per cent of the total legume N.

- Perennial legumes may fix 100 to 200 lb/a/yr
- Annual legumes fix 50 to 100 lb/a/yr.

Small amounts of ammonium can also be released by roots of the legume into the rhizosphere, or the surrounding soil.

Availability of Nitrogen in Subsequent Cropping Systems

Research shows that yields of non-legume crops can increase when following a legume rotation. It is believed that the legume rotation increases the N content of soil, thus making it an effective nutrient management strategy.

However, when the legume is incorporated into the soil, the major benefit of the legume rotation lasts only during the first year following the legume rotation.

The Elements of Complete Plant Nutrition

The following is a brief guideline of the role of essential and beneficial mineral nutrients that are crucial for growth. Eliminate any one of these elements, and plants will display abnormalities of growth, deficiency symptoms, or may not reproduce normally.

Macronutrients

- Nitrogen is a major component of proteins, hormones, chlorophyll, vitamins and enzymes essential for plant life. Nitrogen metabolism is a major factor in stem and leaf growth (vegetative growth). Too much can delay flowering and fruiting. Deficiencies can reduce yields, cause yellowing of the leaves and stunt growth.
- Phosphorus is necessary for seed germination, photosynthesis, protein formation and almost all aspects of growth and metabolism in plants. It is essential for flower and fruit formation. Low pH (<4) results in phosphate being chemically locked up in organic soils. Deficiency symptoms are purple stems and leaves; maturity and growth are retarded. Yields of fruit and flowers are poor. Premature drop of fruits and flowers may often occur. Phosphorus must be applied close to the plant's roots in order for the plant to utilise it. Large applications of phosphorus without adequate levels of zinc can cause a zinc deficiency.

- Potassium is necessary for formation of sugars, starches, carbohydrates, protein synthesis and cell division in roots and other parts of the plant. It helps to adjust water balance, improves stem rigidity and cold hardiness, enhances flavour and colour on fruit and vegetable crops, increases the oil content of fruits and is important for leafy crops. Deficiencies result in low yields, mottled, spotted or curled leaves, scorched or burned look to leaves..
- Sulfur is a structural component of amino acids, proteins, vitamins and enzymes and is essential to produce chlorophyll. It imparts flavour to many vegetables. Deficiencies show as light green leaves. Sulfur is readily lost by leaching from soils and should be applied with a nutrient formula. Some water supplies may contain Sulfur.
- Magnesium is a critical structural component of the chlorophyll molecule and is necessary for functioning of plant enzymes to produce carbohydrates, sugars and fats. It is used for fruit and nut formation and essential for germination of seeds. Deficient plants appear chlorotic, show yellowing between veins of older leaves; leaves may droop. Magnesium is leached by watering and must be supplied when feeding. It can be applied as a foliar spray to correct deficiencies.
- Calcium activates enzymes, is a structural component of cell walls, influences water movement in cells and is necessary for cell growth and division. Some plants must have calcium to take up nitrogen and other minerals. Calcium is easily leached. Calcium, once deposited in plant tissue, is immobile (non–translocatable) so there must be a constant supply for growth. Deficiency causes stunting of new growth in stems, flowers and roots. Symptoms range from distorted new growth to black spots on leaves and fruit. Yellow leaf margins may also appear.
- Carbon forms the backbone of many plants biomolecules, including starches and cellulose. Carbon is fixed through photosynthesis from the carbon dioxide in the air and is a part of the carbohydrates that store energy in the plant.
- Hydrogen also is necessary for building sugars and building the plant. It is obtained almost entirely from water. Hydrogen ions are imperative for a proton gradient to help drive the electron transport chain in photosynthesis and for respiration.
- Oxygen is necessary for cellular respiration. Cellular respiration is the process of generating energy–rich adenosine triphosphate (ATP) via the consumption of sugars made in photosynthesis. Plants produce oxygen gas during photosynthesis to produce glucose but then require oxygen to undergo aerobic cellular respiration and break down this glucose and produce ATP.

Micronutrients:

- Iron is necessary for many enzyme functions and as a catalyst for the synthesis of chlorophyll. It is essential for the young growing parts of plants. Deficiencies are pale leaf colour of young leaves followed by yellowing of leaves and large veins. Iron is lost by leaching and is held in the lower portions of the soil structure. Under conditions of high pH (alkaline) iron is rendered unavailable to plants. When soils are alkaline, iron may be abundant but unavailable. Applications of an acid nutrient formula containing iron chelates, held in soluble form, should correct the problem.
- Manganese is involved in enzyme activity for photosynthesis, respiration, and nitrogen metabolism. Deficiency in young leaves may show a network of green veins on a light green background similar to an iron deficiency.
- In the advanced stages the light green parts become white, and leaves are shed. Brownish, black, or grayish spots may appear next to the veins. In neutral or alkaline soils plants often show deficiency symptoms. In highly acid soils, manganese may be available to the extent that it results in toxicity.
- Boron is necessary for cell wall formation, membrane integrity, calcium uptake and may aid in the translocation of sugars. oron affects at least 16 functions in plants. These functions include flowering, pollen germination, fruiting, cell division, water relationships and the movement of hormones.
- Boron must be available throughout the life of the plant. It is not translocated and is easily leached from soils. Deficiencies kill terminal buds leaving a rosette effect on the plant. Leaves are thick, curled and brittle. Fruits, tubers and roots are discoloured, cracked and flecked with brown spots.
- Zinc is a component of enzymes or a functional cofactor of a large number of enzymes including auxins (plant growth hormones). It is essential to carbohydrate metabolism, protein synthesis and internodal elongation (stem growth). Deficient plants have mottled leaves with irregular chlorotic areas.
- Zinc deficiency leads to iron deficiency causing similar symptoms. Deficiency occurs on eroded soils and is least available at a pH range of 5.5—7.0. Lowering the pH can render zinc more available to the point of toxicity.
- Copper is concentrated in roots of plants and plays a part in nitrogen metabolism. It is a component of several enzymes and may be part of the enzyme systems that use carbohydrates and proteins. Deficiencies cause die back of the shoot tips, and terminal leaves

develop brown spots. Copper is bound tightly in organic matter and may be deficient in highly organic soils. It is not readily lost from soil but may often be unavailable. Too much copper can cause toxicity.

- Molybdenum is a structural component of the enzyme that reduces nitrates to ammonia. Without it, the synthesis of proteins is blocked and plant growth ceases. Root nodule (nitrogen fixing) bacteria also require it. Seeds may not form completely, and nitrogen deficiency may occur if plants are lacking molybdenum. Deficiency signs are pale green leaves with rolled or cupped margins.
- Chlorine is involved in osmosis, the ionic balance necessary for plants to take up mineral elements and in photosynthesis. Deficiency symptoms include wilting, stubby roots, chlorosis (yellowing) and bronsing. Odours in some plants may be decreased. Chloride, the ionic form of chlorine used by plants, is usually found in soluble forms and is lost by leaching. Some plants may show signs of toxicity if levels are too high.
- Nickel has just recently won the status as an essential trace element for plants according to the Agricultural Research Service Plant, Soil and Nutrition Laboratory in Ithaca, NY. It is required for the enzyme urease to break down urea to liberate the nitrogen into a usable form for plants. Nickel is required for iron absorption. Seeds need nickel in order to germinate. Plants grown without additional nickel will gradually reach a deficient level at about the time they mature and begin reproductive growth. If nickel is deficient plants may fail to produce viable seeds.
- Sodium is involved in osmotic (water movement) and ionic balance in plants.
- Cobalt is required for nitrogen fixation in legumes and in root nodules of non-legumes. The demand for cobalt is much higher for nitrogen fixation than for ammonium nutrition. Deficient levels could result in nitrogen deficiency symptoms.
- Silicon is found as a component of cell walls. Plants with supplies of soluble silicon produce stronger, tougher cell walls making them a mechanical barrier to piercing and sucking insects. This significantly enhances plant heat and drought tolerance. Foliar sprays of silicon have also shown benefits reducing populations of aphids on field crops. Tests have also found that silicon can be deposited by the plants at the site of infection by fungus to combat the penetration of the cell walls by the attacking fungus. Improved leaf erectness, stem strength and prevention or depression of iron and manganese toxicity have all been noted as effects from silicon. Silicon has not been determined essential for all plants but may be beneficial for many.

Minor Elements in Plant Nutrition

Copper

Copper has long been known as an element toxic to plants even when present in what may be considered high dilution. One part in 10 million of water is toxic to fresh-water algae and copper sulphate is commonly used in the elimination of water weeds from tanks, etc. This element is receiving more and more attention as a minor element. Its increasing importance arises from the fact that copper deficiency causes pathological symptoms in animals as well as in plants. In the higher plants a modicum of copper is essential, and many soils, often those of a peaty character, fail to provide the very small amount necessary.

The addition of to of one part per million in water culture, or about 1 per cent to 3 Ib. of copper per acre, seems to eliminate all trouble in this respect. The element is often supplied in the form of the sulphate or pyrites: the amount of either of these to be sown depends on their copper content. The mechanism through which the copper acts is not known. That plants deficient in copper are often lacking in chlorophyll is not without importance, for copper deficiency in the diet of infants is often the cause of some of the anaemias that they suffer from.

The red pigment of blood and the green pigment of the plant are related in very many respects. Where copper is in too great supply it has been shown that iron acts as a "depoisoner." Deficiency of zinc, too, causes pathological symptoms in the plant. It is thought that the curative action of some fungicidal zinc sprays is more connected with the action of the metal on the physiology of the host that its direct effect on the disease organism.

Cobalt

Soils deficient in cobalt produce herbage of normal appearance, but the animals fed on the produce pine and die. The addition of about 2 Ib. cobalt chloride per acre eliminates the trouble.

Boron

Of all the minor elements probably none has received so much attention as boron. It is deficient in quantity in many soils all over the world. Symptoms of disfunction following from apparent boron deficiency seem to be connected with calcium supply, and show fundamentally in reduction or stoppage of activity in the meristematic regions. Further, plants showing symptoms of boron deficiency are found usually in calcareous soils or those which have been heavily limed.

Boron does not seem to be necessary to animals, so that the chief interest in this minor nutrient resides in the pathological condition it may set up in the plant.

Manganese

Manganese deficiency symptoms, too, are most commonly found in plants growing on soils well supplied with lime. Except at high dilution manganese is toxic to plants, yet when presented in too low a concentration metabolic upsets result. These are definitely connected with lowered efficiency of the respiratory function. Indeed manganese seems to be necessary for the whole oxidation/reduction mechanism. Plants offered a nitrogen supply in the form of ammonia compounds are benefited by extremely small additions of manganese to the nutrient medium.

This is especially so when the oxygen supply is limited. Additions of manganese seem to benefit the legumes more than other cultivated plants. A sufficiency of manganese in the food is necessary for animals.

Molybdenum

Molybdenum was fairly early shown to be necessary for the nutrition of many fungi, but it is now thought that higher plants use it. Certain soils, typically those lying on the Lower Lias formation in south-east England, contain about 20 to 100 parts per million of this element, and the herbage produced on them may contain about 2 to 3 parts per million. When grasing on such a pasture ruminant animals scour, pine, and eventually die. Such pastures are known as "teart." Clovers seem to absorb more molybdenum than grasses, hence increase of these valuable plants intensifies the bad effect.

It would seem that the "availability" of molybdenum is controlled by the acidity of the soil. In neutral or alkaline soils the availability is high; in acid conditions low. Addition of lime to the soil does not render teart pastures healthy, but rather the reverse.

Also, the clovers being encouraged by the lime become more prolific, and because of their high molybdenum content they raise the content of the herbage as a whole. There is some evidence that copper may be antagonistic to molybdenum and be useful in treating affected animals or in improving teart pastures. Other elements affect the plant and through the plant, animals. Enough has been said to elucidate the principles involved, and for fuller details of each case reference should be made to the very voluminous literature which has already grown up.

Compounds of Elements

Selenium

It will be convenient to discuss -one of the non-essential elements here as it is intimately related to sulphur. This is selenium, which is of special interest because it is often absorbed by plants in a quantity sufficient to render them toxic to animals while themselves remaining unaffected.

The element is comparatively rare, but traces of its compounds in the soil result in herbage, grain, etc., producing all the symptoms of the dreaded alkali disease when fed to animals. Man, too, is subject to selenium poisoning. There is no doubt that plants "heap up"selenium, and that the accumulation is due in large part to the metal replacing the sulphur in certain of the amino-acids of the protein molecule. The process of "heaping up"need not go very far for the plant to become toxic to animals. Rats die when fed on perfectly normal-appearing plants containing 2 or 3 parts per miJlion of selenium. On the other hand, "heaping-up"must go very far before the plants show symptoms. Fifteen parts per million in the plant produces chlorosis, and with greater dosage the leaves and roots develop a pink colouration.

Fortunately selenium and sulphur are antagonistic. Addition of sulphur to the soil substantially reduces the amount of selenium subsequently found in the tissues of the plant. By applications of gypsum, selenium-containing soil can be made to produce crops which are non-toxic to animals.

Selenium-tolerant Plants

Some plants not only tolerate this element so toxic to animals but actually require it for successful growth. *Astragalus racemosus* and *A. pattersonia* are reported as thriving in a medium containing 27 parts of selenium per million. This is about ten times the amount tolerated by wheat or buckwheat.

Furthermore, in these plants addition of sulphur to the soil, while resulting in considerable reduction of the selenium content, does not bring it down to a level likely to be tolerated by grasing animals. Plants which occur naturally on seleniferous soils and nowhere else are now becoming known. It is expected that delimitation of seleniferous areas by mapping the occurrence and distribution of these "indicator"plants will be comparatively simple, and so eliminate much chemical work on soil analysis.

Selenium as an Insecticide

An interesting development of these discoveries regarding selenium is that crops can be protected from biting or sucking insects if selenium is added to the soil in which the plants grow. Insects die after feeding on the seleniferous juice. This can be made use of without detriment to man or his animals only when the harvested produce is not consumed as food, *e.g.* timber and fibres, or where it is food but not protein, *e.g.* sugar. A slight exception to this is seen in cacao where the very young plant is subject to attacks from insect pests.

The juvenile plantlets are supplied for a short period with selenium, after which they are transplanted to fresh untreated land. By the time the plants are adult and bearing a crop the small amount of this element which protected the small plant is dispersed through a large plant, and the small percentage then present is not toxic in the product.

Phosphorus

Phosphorus is of fundamental importance in the economy of the plant, involved as it is in protein formation, as well as in respiration and other functions. The utilisation of nitrogen by the plant is interfered with if phosphorus is in deficient supply. This may be due directly to insufficiency of phosphate for the formation of protein. There is evidence, however, that the action lies deeper, for when phosphates are deficient the enzyme responsible for the initial steps in nitrate utilisation is either absent or inoperative. Lack of phosphate seems, then, to "bottleneck "the whole of the synthesis in the plant.

Mono- and di-saccharides "bank-up" amino and amide nitrogen accumulate, and eventually cell-division slows down and ceases. The slowing down of cell-division may be a reflex of deficiency of necessary phosphates for protoplasm formation, but it is also known that When phosphorus supply is inadequate the fibres which "pull" the split chromosomes apart in mitosis are not formed.

A concomitant of this whole imbalance of cell function is cessation of all phenomena requiring active cell-division, *e.g.* growth of the axis and branching. This is particularly apparent in cereals where the result is that tillering is depressed. The' converse of these effects is seen when phosphate is supplied liberally, for there is then more active branching of the root and shoot, while the plant tends to flower and seed not only earlier but more profusely.

Calcium

The predominant role of calcium in agriculture is not wholly botanical, but arises from the use of lime as a soil ameliorant and as a reducer of soil acidity. The element, however, is taken into the plant in quite substantial quantities, probably as the metallic ion. In the tissues it neutralises acids arising from many metabolic activities, and crystals of the calcium salts of oxalic and other organic acids commonly appear in cells. One of the principal roles of calcium in the plant seems to be involved in some fine balance with boron, and possibly other elements.

What the relationship is still remains obscure. Certainly deficiencies of both boron and calcium are very similar in the symptoms they produce. This is especially so in the failure of nornytLetevelopment in meristematic regions. Calcium is found in great quantities in leaves, and does not migrate much from an old organ to a young one as the plant develops. Calcium is an essential ingredient of the structure of the cell wall, or at least of the middle lamella.

Nitrogen

Let us now turn to the special characteristics of each element in connection with intake. It was believed at one time that the plant used nitrogen only as the nitrate ion NO_3, and that ammonia and other forms had to be converted to this radicle in the soil before entry or utilisation was possible. It is now known that in the field ammonia is used satisfactorily by some plants. Whether any particular kind of plant can make use of ammonia or not would seem to depend on whether it can "step up"this form of nitrogen quickly to higher combinations such as an amide. Apparently ammonia cannot be stored as such in the cells.

The rate of intake of ammonia seems to be faster than the intake of nitrate—at least in such plants as the cotton seedling, oats, and buckwheat. Whether ammonia or nitrate is the more suiform seems to be governed by at least two factors, the nature of the plant and the reaction of the soil.

An acid soil favours the use of nitrate. The successful use of nitrite (NO_2) by plants has been reported, but is of little agricultural interest. The chief use of nitrogen in the plant is in the formation of compounds such as amino-acids, which enter into the molecular structure of the proteins. Since proteins are the basis of protoplasm, reduced nitrogen supply soon shows itself in reduction of protoplasm formation. This puts a restriction on practically all functions, and particularly on new cell formation, so that growth slows and may cease.

Sulphur

Sulphur is used by the plant only when offered in the form of a sulphate (SO_4); most other forms are toxic or not usable. The sulphur enters into certain amino-acids which are essential in the formation of many proteins. Many plants can maintain apparently normal growth and appearance while suffering from partial lack of sulphur. The tissues are, however, deficient in the element.

A peculiar case of sulphur deficiency occurs in some grassland areas only in spring when growth of the grass is very rapid. The sulphur intake at this time is not sufficient to supply what is necessary for normal protein building, but the plant goes on growing, using such sulphur as it has very economically, or improvising in some other way. A mass of herbage deficient in sulphur is produced. Later in the year, when growth slows down, the sulphur supply is sufficient, so that the maladjustment corrects itself, and normal protein formation follows.

Break in Wool

As wool protein contains much sulphur, sheep grasing in such an area during the spring develop a peculiar characteristic of the wool. When a lock of the wool is sharply pulled, each fibre snaps at the same point, and the broken tuft shows a clean break instead of the irregular break obtained with normal wool. When wool fibres of this type are seen under the microscope an identical part of each fibre is seen to be of reduced diameter. This is the weak point, and is where the break takes place.

Working from known development rates for wool fibre, it appears that the weak part is produced while the sheep is feeding on the spring flush of grass, which is deficient in sulphur. The condition is remedied by feeding during the deficiency period a "lick" of hydrolised wool, which supplies organic sulphur.

Potassium

Potassium in plant nutrition offers something of a paradox in that commonly there is a considerable quantity of potash in the soil with only a small proportion soluble, while in the plant practically the whole of the potassium remains soluble in the cell sap. When a plant tissue is mashed up the potassium may be easily leached out. Despite its appearance in this simple form potassium has fundamental roles to perform. Potassium is concerned with the synthesis of carbohydrates, and probably with their translocation.

This connection between potassium and carbohydrates is clearly seen in crops which normally accumulate large reserves of starch or sugar. These crops when grown in a deficiency of sunlight produce less reserve food (lower crop yield), but when generously supplied with potash the production is greatly increased.

It would appear that potassium can "replace"sunlight. Potassium is moved out of adult tissues to areas of meristematic activity. Potash confers on plants some degree of resistance to various diseases. The clovers and their allies, though not "carbohydrate"plants, gain greater benefit from increased potash supplies than do grasses, and hence in pastures with a grass/clover sward addition of potash tends to strengthen clovers rather than the grasses.

Magnesium

Magnesium is an essential ingredient for the formation of chlorophyll, and indeed is the only metallic constituent of this all-important green pigment of plants. As the plant passes to final maturity and the leaves change colour, magnesium is largely transferred from the vegetative body to the seed.

In the plant another of its functions is connected with its use along with phosphate in the formation of those special proteins which form the cell nucleus (the nucleo-proteins).

Magnesium, too, appears to be concerned in oil metabolism. The amount of magnesium required by the plant is not large and, when too great quantity is available in the soil functional disturbances develop in the plant.

Iron

Iron is not an essential element in the sense that the term has so far been used, for only a little is required. Neither is it regarded as a minor element, but occupies an intermediate position in the classification. Iron seems to take no structural part in the plant, yet its presence is essential for the formation of chlorophyll. Though necessary for the formation of chlorophyll, it does not enter into the molecule of that substance. Iron is not mobile inside the plant. This is easily shown by first growing plants for a time in a solution supplying iron when the shoot is fully green.

If the experimental plants are then transferred to a culture medium deficient in iron, the new leaves produced in further growth are white due to a lack of iron in them inhibiting formation of chlorophyll.

Total absence of iron from soils is not so common as total or nearly total non-availability. Thus it is that many soils rich in iron produce plants showing all the symptoms of iron deficiency.

The Calcium/Phosphorus Ratio

The importance in the diet of animals of adequate quantities of calcium and phosphorus individually, and of the ratio of these to each other, needs no stressing here. The data offered in the general account of salt intake should be again examined from this aspect of the quality of food.

Soil Compaction and Building

Soil compaction is defined as the method of mechanically increasing the density of soil. In construction, this is a significant part of the building process.

If performed improperly, settlement of the soil could occur and result in unnecessary maintenance costs or structure failure. Almost all types of building sites and construction projects utilize mechanical compaction techniques

Soil density is used almost exclusively by the transportation industry to specify, estimate, measure, and control soil compaction. Soil density can be easily determined via weight and volume measurements. The objective of compaction is to stabilize soils and improve their engineering properties. Source: US Dept. of Transportation

There are five principle reasons to compact soil:-

- Increases load-bearing capacity-
- Prevents soil settlement and frost damage-
- Provides stability-
- Reduces water seepage, swelling and contraction-
- Reduces settling of soil-source Soil Compaction Handbook.

Porosity

Bulk Density is an indirect measure of soil pore space. In fact, there is a formula: 100-((B.D. ÷ Particle Density) X 100)=% pore space or POROSITY. Under field conditions, pore spaces are occupied at all times by air and water. "Tortuous pathways" best describes soil pores. Porosity, when expressed as a percent, is the same thing as percent pore space. Soil particles have irregular shapes, and thus the spaces or pores between them vary irregularly in size, shape and direction. Sandy soils have large continuous pores, while clays have small pores which transmit water slowly. Clays, however, contain more pore space than sandy soils, because of the pores inside of the soil peds. To growing plants, pore sizes are of more importance than total pore space.

Particle Density

In the laboratory you will determine porosity and particle density for the sandy soil used previously. Particle density will also be determined for this soil. Particle density only takes into account the volume occupied by the solid particles. It excludes the volume occupied by air and water.

Since a large portion of most soils is composed of particles derived from quartz minerals, the particle density of most soils is near 2.6 g/cc, which is the density of quartz. Variations in the particle density are due to the presence of heavier minerals like iron oxides or lighter organic components.

Significance of Porosity

Above, a weakly developed soil is on the left and a well developed soil is on the right. Where clay accumulation is minimal, the macro pore space is about the same with increasing depth.

However, where clay has moved into pores, the amount of macro pores and the rate of water movement both decrease. Soils with clay films will have slow water flow through the Bt horizon.

Some soils are difficult to aerate by plowing due to the turf cover. For lawns, playgrounds, or football fields, aeration is accomplished by making small holes in the turf with solid tines or using hollow tines to pull out a small core. Later the holes can filled with sand to promote air exchange. This is called *aerifying* and *topdressing*.

Many home lawns are compacted because of the construction activities when the home was built or due to people walking or riding on the turf when the soil is wet. Aerification will improve the turf growth on these lawns.

Aeration Equipment=Tines & John Deere

Soil Temperature Regimes

The temperature of a given soil at a given time is dependent upon the gains and losses of heat energy. Generally, dark surfaces will absorb more heat than light surfaces. However, the amount of water in the soil is an important factor. Losses of the absorbed heat are by radiation back into the atmosphere as long-wave radiation, heating the air and cooling the soil.

These soils warm quickly in the spring due to their dark surface.

Soil temperature regimes are used to classify soils. They are defined according to the average annual soil temperature in the root zone. The use of soils for agriculture and forestry is closely related to soil temperature, due to the specific requirements of plants. Over most of the earth, daily soil temperatures below 50 cm deep seldom change. To approximate the mean annual soil temperature, 2 ° F is added to the mean annual air temperature. If we do this for the annual temperature map of Minnesota and follow the 45°F line, which would be 47° soil temperature, the state is divided along a line from the Twin Cities to the South Dakota border, and soils are frigid to the north and mesic to the south.

Soil Bulk Density Determination

Soil weight is referred to as soil bulk density. Density is the mass of material contained within a given volume. This is the idea that a given size of box may be heavy or light, depending upon what kind of material it contains. If the box were filled with wood, it would be light when compared to having it filled with lead. The weight of water is the reference for density measurements: 1 gram of water=1 cubic centimeter (cc), & 1 cc water=1 ml. & 1 cubic foot of water=62.4 lbs.

The bulk density of a soil is the mass of dry soil per unit of "bulk" (total volume of soil or soil particles & pore space).

B.D. = mass of oven dry soil (grams) ÷ total volume of soil (cm^3)=grams/cc

The bulk density takes into account the total soil volume (the space occupied by the solid particles plus the space occupied by the air of the pores or pore space). To determine the mass of the soil—since air does not have any significant weight—we can just weigh the oven dry soil on a balance. The volume of the soil can be determined by pouring the soil into a graduated cylinder and measuring the volume that it occupies.

The problem with this procedure is that structural aggregates may be crushed or compacted once they are removed from the soil and placed into the cylinder. A better procedure is one in which the aggregates could be removed from the soil and frozen exactly the way they were in the soil. Plastic fixatives enable us to do this.

A large aggregate has been removed from the soil and is being fixed by dipping it into a saran solution. This will make the clod impervious to water. The weight of our clump of soil can be obtained by hanging it on the balance or placing it on the metal weighing tray.

The volume of the clod also needs to be determined. This can be obtained by weighing the sample again, only this time the sample will be in water. The weight of the sample now will be less, since the sample will be buoyed up by the amount of water it is displacing.

Archimedes most famous theorem gives the weight of a body immersed in a liquid, called Archimedes' principal. If you want to know more about the mathematician who determined that this procedure would work, go to Archimedes.

Thus, by subtracting the weight of the clod in water from its weight in air, we obtain the weight of the water displaced by the clod, which equals the volume of water displaced by the clod (because 1 cm^3 of water=1 gm. of water) and is equal to the volume of the soil clod.

Clod Method of BD Determination =Mass of Clod ÷ Volume of water displaced

Volume of water displaced=weight of water or the (gms clod in air)-(gms clod in water)=(grams or volume of water displaced)

Cylinder method of BD Determination=Mass of oven dry soil (gms) ÷ total volume of soil (cm^3)

Another method of determining bulk density is by using soil cores. These are obtained with a probe that forces a cylinder into the soil. The soil can be removed from this cylinder and cut into sections and placed in another cylinder for transportation back to the laboratory. Care needs to be taken to avoid compacting the soil during and after obtaining the core.

BD of soil (using core method)={(mass of soil + mass of core)-(mass of core} ÷ (volume of core)

volume of core=Pi (3.1416) $r^2 \times h$

r=radius of core; h=the height of core.

For the cores in the lab, r=2.45cm (½ of Diameter); h=5 cm

$V=3.1416 \times 2.45^2 \times 5=94.3$ cm^3;

Bulk density is a fairly easy determination that will yield information about the soil that will be significant for determining the soil's potential for plant growth or a building foundation. Remember it is the oven dry mass/volume soil. or g/cc.

Significance of Bulk Density

The bulk density of the soil will play an important role in determining if the soil has the physical characteristics necessary for plant growth, building foundations or other uses.

From the laboratory investigations you will obtain very high bulk densities for the Bt and sandy soil. The weight of the soil is important if you are going to be lifting it or hauling it long distances.

One of the main reason for sod farms to be on organic soils is to reduce the cost of transportation because the organic soil is so much lighter than mineral soils. The organic sod is easier to handle. Most sod farms in Minnesota are on the peat soils north of the Twin Cities.

Sometimes we are interested in the weight of an acre of soil for erosion comparison purposes. Erosion of 5 tons per acre sounds like a lot. However the weight of an acre furrow slice on average is 2,000,000 lbs., so 5 tons per acre seems like a small amount. Five tons per acre is only about 34 thousandths of an inch thick (.034 inches). However, many areas have lost 10 to 100 times that amount. (Note: an acre furrow slice, or AFS, is a three-dimensional volume of soil which is one acre in area and 7 inches deep.)

Soil Factors Impacting Tillage

Another factor similar to carrying soil is its potential to be moved short distances, such as in plowingor rototilling. The soil that is heavier will be more difficult to move. However, another factor is the ability of the soil to stick together, or the soil's consistence. Clay soil is noted for its stickiness and large energy requirements in tillage.

Farmers refer to it as "heavy," but they really mean it is difficult to plow, not that it has a high bulk density; clay soils generally will have a lower bulk density than sandy soils (clay=1.3 g/cc vs sand=1.6g/cc). Sandy soils have a higher bulk density, but are easier to plow since they have weaker consistence. Thus they are often referred to as "light soils." The lower B.D. of clays is due to their better aggregation.

Rototilling the soil reduces the bulk density by "fluffing" the soil. For a look at tillage implements go to Tillage Implements

Soil Consistence

Soil consistence is the soil's ability to cohere or stick together. The soil's consistence may be evaluated at three moisture conditions: air dry, moist, and wet. Moist consistence is evaluated by placing the soil between the thumb and forefinger and gently applying pressure.

The ease with which a ped can be crushed determines the consistency.

Terms commonly used to describe moist consistence are:

Loose-Non-coherent when dry or moist; does not hold together in a mass.

Friable-When moist, crushes easily under gentle pressure between thumb and forefinger and can be pressed together into a lump.

Firm-When moist, crushes under moderate pressure between thumb and forefinger, but resistance is distinctly noticeable.

Plastic-When wet, readily deformed by moderate pressure but can be pressed into a lump; will form a "wire" when rolled between thumb and forefinger.

Sticky-When wet, adheres to other material and tends to stretch somewhat and pull apart rather than to pull free from other material.

Hard-When dry, moderately resistant to pressure; can be broken with difficulty between thumb and forefinger.

Soft-When dry, breaks into powder or individual grains under very slight pressure.

Cemented-Hard; little affected by moistening.

In air dry conditions, the resistance to rupturing when rubbed is measured.

At intermediate moisture content, the soil's resistance to shearing forces by thumb and finger is noted. In the wet condition, its plasticity—ability to be molded and stickiness—are measured.

One significance of the B.D. of a soil is the amount of surface area a soil has. The more surface area, the more ability to retain water and nutrients. Notice that the two soils pictured below weigh the same, but are significantly different in surface area. Which soil has the higher B.D.?

Compaction, Porosity and Soil Temperature

Soil Compaction

Soil compaction works something like the soil particles seen below.

When the soil is randomly fluffed, as on the left, the amount of pore space will be 45 to 55%. When we push the soil particles closer together, we remove the pore space, and the bulk density will increase. Compaction increases the bulk density by reducing the pore space.

Compaction can also result in the changing of the proportion of pore sizes. Notice in the compaction diagram that compacted soils not only have a lower total pore space (resulting in higher bulk density), but also less macro pores and more micro pores.

This will often result in excess water retention.

When we walk on the soil we compact it. Eventually the compaction prevents the growth of plants. This trail TRAIL behind the Soils Building shows the effect of compaction and samples from the soil will be used in the lab to determine the amount of compaction.

Compaction changes the amount and size of pores.

Deep plowing can reduce compaction below the plow zone.

C.O.L.E.

One of the important engineering properties of soils is related to the clay content and consistency and is called the coefficient of linear extensibility or C.O.L.E. It is a measure of the shrink-swell potential, or the volume change of the soil with changes in moisture content. If the C.O.L.E. exceeds.09, significant shrink-swell potential can be expected.

Soils with high C.O.L.E. values can cave in basements, displace walls, and break gas or water pipes. Knowing which soils have high C.O.L.E. values is an important step in selecting sites for construction activities. Compacting soils reduces the ability of soils to take on water and reduces their shrink/swell potential.

C.O.L.E.=(Length Moist ÷ Length Dry)-1

Soil Analysis: A key to Soil Nutrient Mangement

High yields of top-quality crops require an abundant supply of 16 essential nutrient elements. In addition to providing a place for crops to grow, soil is the source for most of the essential nutrients required by the crop. Our soil resource can be compared to a bank where continued withdrawal without repayment cannot continue indefinitely. As nutrients are removed by one crop and not replaced for subsequent crop production, yields will decrease accordingly.

Accurate accounting of nutrient removal and replacement, crop production statistics, and soil analysis results will help the producer manage fertilizer applications.

A soil analysis is used to determine the level of nutrients found in a soil sample. As such, it can only be as accurate as the sample taken in a particular field.

The results of a soil analysis provide the agricultural producer with an estimate of the amount of fertilizer nutrients needed to supplement those in the soil. Applying the appropriate type and amount of needed fertilizer will give the agricultural a more reasonable chance to obtain the desired crop yield.

Objectives of Soil Analysis

- To provide an index of nutrient availability or supply in a given soil. The soil extract is designed to evaluate a portion of the nutrients from the same "pool" used by the plant.
- To predict the probability of obtaining a profitable response to fertilizer application. Low analysis soils may not always respond to fertilizer applications due to other limiting factors. However, the probability of a response is greater than on a high analysis soil.
- To provide a basis for fertilizer recommendations for a given crop.
- To evaluate the fertility status of the soil and plan a nutrient management programme.

Chemical analysis of plant composition indicates chemicals or elements present in a crop at maturity or when it is harvested. For example, 1,250 lb of lint cotton contains approximately 125 lb of nitrogen (N), 20 lb of phosphorus (P), and 75 lb of potassium (K).

The essential question in fertilization is, "How much nutrient must be added to the soil as fertilizer for a given amount to be taken up by the growing plant?" The crop utilizes only a portion of the available nutrients in the soil. This means that more nutrients must be present than are removed by the crop. The amount added varies according to the level already present in the soil and the crop's need for the nutrient involved. The soil analysis is the starting point, since it measures the level or content presently in the soil.

The soil analysis along with the information provided in the information sheet, is interpreted and reported in terms of the nutrients needed to supplement those in the soil. With this information, producers can add sufficient nutrients for the correct balance to obtain high yields.

Limiting Factors

Crop yields are determined by a variety of factors including crop variety selection, available moisture, soil fertility, crop adaptation to the area, and the presence of diseases, insects, and weeds. The soil analysis and its interpretation deal only with the fertility level (plant nutrients) of the soil. Recommended fertilizer will provide sufficient nutrients for the best possible yields. Other factors of production or management may still cause low yields, even though nutrients are adequate.

Carryover

If yields are only partial in relation to a large amount of fertilizer applied, many of the nutrients are carried over for use by the next crop. It is this carryover, or residual effect, from one year to the next that makes heavy fertilizer applications practical in the face of other limits to yield.

Yields to Expect

A certain fertilizer application cannot be expected to produce a specific yield such as two bales of cotton or nine tons of hay. It is more realistic to assume that a balanced fertilizer programme assures that nutrients are not the limiting factor in yields obtained. Research has shown that producers who use a balanced fertilizer programme obtain consistently better yields than those who don't.

The Soil Analysis Report

After the soil is analyzed, fertility recommendations are made based on amounts of actual nutrients in the soil, not on the amount of any particular fertilizer or mixture. For example, if 100 lb of N were recommended, that amount could be supplied by approximately 300 lb of ammonium nitrate (33%N), 220 lb of urea (45%N), or 120 lb of anhydrous ammonia (82%N). Likewise, a recommendation of 60 lb of P205 per acre could be added as 133 lb of 45% triple superphosphate.

Fertilizer Labeling

Nitrogen is expressed on the elemental basis as "total nitrogen" (N). Phosphorus is expressed on the oxide basis as "available phosphoric acid" (P205). Potassium is expressed as "soluble potash" or potassium oxide (K20).

In reality, there is no P205 or K20 in fertilizers. Phosphorus exists most commonly as monocalcium phosphate, but also occurs as other calcium or ammonium phosphates. Potassium is ordinarily in the form of potassium chloride or sulfate. Furthermore, P205 and K20 are not absorbed by plants. Plant roots absorb most of their phosphorus in the form of orthophosphate ions, H2P04-, and most of their potassium as potassium ions, K+. For these reasons, the elemental expression (N-P-K) is used in all of the recent research publications. Conversions from one form of P and K to another can be made using the following formulas.

$$\%P=\%P205 \times 0.437 \quad \%K=\%K20 \times 0.826$$
$$\%P205=\%P \times 2.29 \quad \%K20=\%K \times 1.21$$

Interpretation of the Soil Analysis Report

The soil analysis report contains two parts: characterization and fertility status of the soil, and fertility recommendations. Soil characterization (pH, texture, percent exchangeable sodium, percent organic matter, and salinity expressed as electrical conductivity) is explained in the report. The fertility status is reported as nutrients available to the plant. The second part, fertility recommendation, contains the suggested amounts of fertilizer to apply. These amounts are based on the crop requirements, management practices affecting the crop (as shown in the information sheet), the present fertility level of the soil, and the yield goal desired by the producer. Special notification is given if the tests indicate that a salt or sodium hazard exists or if the information provided shows any other specific problems.

Soil amendments or treatments to reduce a sodium or salt hazard will be recommended if requested. In general, application of gypsum is suggested for reducing a sodium hazard, and leaching is recommended in most cases to lower salt content in the soil. Gypsum or leaching requirements are calculated and reported if requested.

Where to Get Soil Analyzed

There are many soil testing laboratories in New Mexico, Texas, Colorado, and Arizona. Basic soil testing packages vary in price and number of analyses. Many labs are participating in the Western Region Soil Testing Proficiency Programme. Programme participants share identical soils and compare results quarterly. This process assures the clients that the lab is striving for consistency and accuracy in lab analyses. Recommendations will undoubtedly vary from lab to lab. Often the best recommendation will come from the local Extension service.

The choice of labs is at the client's discretion but should be based on report readability, result accuracy, turn-around time, and cost factors. New Mexico specialists can assist with many questions regarding plant health. Remember, a soil analysis is only as good as the soil sample taken.

Soil Analysis: Key to Nutrient Management Planning

Soil provides a reservoir of nutrients required by crops and also therefore for animals but not necessarily at optimum levels of immediate availability to plants. The purpose of soil analysis is to assess the adequacy, surplus or deficiency of available nutrients for crop growth and to monitor change brought about by farming practices.

This information is needed for optimum production, to avoid transferring undesirable levels of some nutrients into the environment and to ensure a suitable nutrient content in crop products.

Farm assurance schemes, buyer's protocols and codes of practice are increasingly demanding more accurate fertiliser recommendations which must depend on the nutrient-supplying capacity of the soil. Regular soil analysis should be undertaken as a vital part of good management practice.

Soil Conservation

Soil conservation is a set of management strategies for prevention of soil being eroded from the Earth's surface or becoming chemically altered by overuse, acidification, salinization or other chemical soil contamination.

It is a component of environmental soil science. Decisions regarding appropriate crop rotation, cover crops, and planted windbreaks are central to the ability of surface soils to retain their integrity, both with respect to erosive forces and chemical change from nutrient depletion.

Crop rotation is simply the conventional alternation of crops on a given field, so that nutrient depletion is avoided from repetitive chemical uptake/ deposition of single crop growth.

Erosion Prevention

There are also conventional practices that farmers have invoked for centuries. These fall into two main categories: contour farming and terracing, standard methods recommended by the U.S. Natural Resources Conservation Service, whose Code 330 is the common standard. Contour farming was practiced by the ancient Phoenicians, and is known to be effective for slopes between two and ten percent. Contour plowing can increase crop yields from 10 to 50 percent, partially as a result from greater soil retention.

There are many erosion control methods that can be used such as conservation tillage systems and crop rotation. Keyline design is an enhancement of contour farming, where the total watershed properties are taken into account in forming the contour lines. Terracing is the practice of creating benches or nearly level layers on a hillside setting.

Terraced farming is more common on small farms and in underdeveloped countries, since mechanized equipment is difficult to deploy in this setting. Human overpopulation is leading to destruction of tropical forests due to widening practices of slash-and-burn and other methods of subsistence farming necessitated by famines in lesser developed countries. A sequel to the deforestation is typically large scale erosion, loss of soil nutrients and sometimes total desertification.

Perimeter Runoff Control

Trees, shrubs and groundcovers are also effective perimeter treatment for soil erosion prevention, by insuring any surface flows are impeded. A special form of this perimeter or inter-row treatment is the use of a "grassway" that both channels and dissipates runoff through surface friction, impeding surface runoff, and encouraging infiltration of the slowed surface water.

Nutritional and Ecological Requirements of Forest

For its growth and for other life processes, the tree requires food and like other green plants it builds up its food materials from simple substances. The chemical constituents of a tree are chiefly carbon, hydrogen and oxygen, together with some nitrogen and certain minerals such as potash, lime, iron and phosphorus.

The carbon is obtained from the small amount of carbon dioxide gas that is always present in the atmosphere and enters the plant through small openings in the leaf surface known as "stomata." Oxygen and hydrogen are obtained from the soil through the root system in the form of water carrying nitrogen (nitrates) and mineral nutrients in solution.

The synthesizing of these elements to form substances that the tree requires -in particular the combination of carbon, hydrogen and oxygen to form carbohydrates — takes place mainly in the leaves in the presence of the characteristically green matter (chlorophyll) and of sunlight, which supplies the necessary energy. Air, water, chlorophyll and light are therefore essential to the production of the carbohydrates-in particular cellulose and lignin -which form by far the greater portion of the tree. Soil moisture is especially important, as a deficiency of water inhibits photosynthesis. Tree species differ markedly in their optimum water requirements and also in their sensitivity to departures from the optimum.

Tree species also differ greatly in the amount of light they require, or, alternatively, in the degree of shade they will tolerate. It is customary to classify species as shade-enduring or tolerant (of shade) and light-demanding or intolerant (of shade), but for most species the younger the tree and the better the soil quality on which it is growing, the greater its tolerance of shade.

The environmental factors important in tree growth — solar radiation, air temperature, humidity, precipitation and the physical and chemical nature of the soil — are usually spoken of as the factors of site or locality and their total effect with reference to a species as the site quality.

Site quality is evaluated either directly by examination and classification of soil and other factors, or indirectly, as for example, by using ground vegetation as an indicator of soil conditions, or by relating the height of dominant trees to the age of the stand.

Since the bulk of the tree is derived from carbon dioxide and water, a relatively small demand is made on the soil in growing a crop of timber; thus land too poor for profitable cultivation may be capable of growing a crop of timber satisfactorily.

Deeply penetrating roots bring up from the deeper soil strata mineral nutrients that would not otherwise be available; a relatively high proportion of these minerals may be deposited annually on the surface soil through the falling of leaves, twigs and other debris.

The leaves, which play such an important part in the nutrition of the tree, function for only a limited time before dying. Deciduous trees lose all their leaves at the end of the growing season, whereas evergreens lose only a portion of the older leaves each year.

The growth and general well-being of a tree depend primarily on soil and weather conditions. Also of vital importance to the future of each member of the forest community is the intense competition that develops for space and food requirements-a struggle that inevitably kills many of the competitors and brings about significant changes in the environment of the survivors.

Longevity

Under favorable conditions a tree usually grows vigorously until it reaches maturity, after which the growth rate gradually slows down. Overmature trees grow little and show signs of decadence. The normal life expectancy of trees varies widely with the species: some temperate and tropical hardwoods reach maturity in less than 100 years and some softwoods in 100 to 200 years, but many species of both softwoods and hardwoods require 200 to 500 years to mature. On the Pacific coast of North America, ages of 500 to 700 years are common; the giant sequoia in California has been known to live for over 3,000 years, an example of extreme longevity.

Resistance to Damage

Trees, like other organisms, are damaged and destroyed by various agencies, including fire, insects, fungi, grazing animals, industrial fumes and such climatic factors as wind, frost and snow. The capacity to resist such damage depends upon the vitality and the protective devices of the species. A knowledge of the relative resistance of different tree species is of importance in forest management.

Distribution

Apart from the chance results of geological history, the continental distribution of tree species and the limits of tree growth, both in altitude and latitude, depend primarily on heat; the east-west distribution, on the other hand, is more often determined by moisture conditions—the rainfall in the interior of a continent may be too scanty to support tree growth. Thus each species of tree has a range of natural distribution determined primarily by climate. Obviously there is a great deal of overlapping and many species grow under similar conditions — in fact it is common to have a number of species sharing the same area in a mixed stand.

Pure stands of a single species do not commonly occur in the natural forest, though stands approaching this condition may be found where a species is well adapted to extreme conditions of climate or soil moisture, or where fire has brought about a situation particularly favorable to one species. The most heterogeneous and complex forest communities are found in tropical regions of high rainfall.

The distribution of a species within its natural range is determined mainly by soil, topography and competing vegetation, as well as by local variations in climate, which also affect the long-term development of many species: climatic races or strains develop, more or less adapted to local variations of climate within the range as a whole. This fact accounts, in part at least, for the wide variation in results obtained by using seed of other than local origin in establishing forests.

Scots pine, for example, has developed a number of climatic races and experience in both Europe and North America shows that efforts to introduce a "foreign" race of this species without a proper correlation of climatic conditions at seed source and planting site will almost certainly prove unsatisfactory, particularly in the susceptibility of the trees to damage by snow and sleet and the incidence of disease.

This, of course, also holds true for attempts to establish any tree as an exotic outside the boundaries of its natural range. However, where the climatic conditions of two regions are similar, the introduction of tree species from the one to the other may be undertaken successfully, as, for example, in the introduction of conifers from the Pacific coast of North America into Great Britain.

An understanding of the various biological factors discussed above is of fundamental importance in the development of forestry practice, particularly in the fields of silviculture and protection. It must be emphasized, however, that practical forest management must depend also on the knowledge and application of sound business principles and on the development of suitable methods and techniques to be used in collecting, analyzing and organizing the basic data to be obtained from the forest.

Branches of Forestry

As is implied by what has been said above, forestry has a number of subdivisions, each of which is concerned with some particular aspect or phase of the subject. Hence, an understanding of forestry itself is dependent on knowledge of its parts, the extent to which they are interdependent and how best they may be co-ordinated.

Forest Protection

Agencies that seriously damage and may destroy forests include fire, insects, diseases, grazing animals and adverse conditions of exposure and climate. The relative importance of these varies with local conditions.

Fire is perhaps the most spectacular. It frequently destroys not only the forest growth, but also the surface organic layer beneath, thus reducing soil fertility. Fire is most serious in the coniferous forests of temperate regions and in the dry forests of the tropics and subtropics and is of course most difficult to control in heavily forested regions with sparse populations and poor accessibility.

Protection against fire is usually one of the first steps taken in the development of a forestry programme. Appropriate legislation is passed and organized staff and equipment and transportation facilities are provided to detect and suppress fires as they occur. Where the country is heavily and continuously wooded, strips may be cut through the forest to break the continuity of tree growth and delay or prevent the spread of fires. Clear-cut strips of this nature are known as "fireguards." Fire itself is sometimes used as a preventive measure where the annual fall of litter and the dying back of ground vegetation produce large accumulations of inflammable debris. The "early burning" of this material when moisture conditions are such as to prevent too rapid spread of fire will reduce the danger during a dry period later in the season.

Recent important advances in the field of fire protection include the use of aircraft, mainly for detection and of meteorological data to appraise more precisely the existing degree of fire hazard, so that dangerous conditions can be quickly recognized and adequate preventive measures taken.

In some European countries forest fires are rare and such as do occur cause negligible damage. In Norway, for example, forest properties may be insured against fire at a low premium rate. There are a number of reasons for this: *(a)* in general the forests are easily accessible and under intensive management, with annual fellings well distributed in a large number of small scattered holdings; *(b)* the forest consciousness of the people is highly developed; *(c)* the climatic conditions tend to be favorable, with well-distributed summer rains and an absence of prolonged hot, dry periods.

Despite the damage it causes, fire cannot be looked upon solely as an enemy of the forest. Under certain conditions controlled burning may prove a useful tool in securing the regeneration of valuable forest trees; it may prepare a suitable seedbed, eliminate competing vegetation, or assist in the distribution of seed. Certain conifers, for example, have serotinous cones and considerable heat is required to open them and release the seed: these species may reproduce abundantly after fire and the resulting stand is often referred to as a "fire type." Extensive even-aged coniferous forests of the northern temperate region owe their origin to fire, as do many other commercially important forests in various parts of the world.

Insects that damage and destroy trees and the fungi and other organisms that cause tree diseases, are present in all forests. In endemic proportions and in balance with their environment, they are a necessary and sometimes innocuous feature of the forest community.

Even under these normal conditions, however, some insects and diseases may seriously damage commercially valuable trees. While the damage caused by fungi is usually slow in developing and of minor account in young trees, it is cumulative and may be so extensive in overmature trees as to render them valueless.

Destructive organisms may also reach *epidemic* proportions and destroy one or more tree species over wide areas. This is particularly likely to happen in overmature forests or when an organism is moved from its natural environment to new surroundings.

A classical example is the chestnut blight *(Endothia parasitica),* which was brought to North America from Asia about 1900 and killed practically all of the sweet chestnut *(Castanea dentata)* in eastern America within a few years. Instances may also be cited of widespread destruction by forest insects, such as the spruce budworm *(Choristoneura fumiferana)* in the spruce and balsam-fir forests of northern and eastern Canada.

Methods used to control insects and disease include the removal of trees most susceptible to attack, the establishment of new stands containing a mixture of species rather than pure stands and discrimination against species susceptible to an insect or disease known to be present in the locality. For example, white pine *(Pinus strobus),* which is often badly attacked by the pine weevil (Pissodes strobi) and the blister rust *(Cronartium ribicola),* should not be encouraged where these pests are commonly found, unless some effective means of control is available.

Other methods of controlling forest pests include the introduction of parasites and predators, the use of chemical sprays and the development through selection and crossbreeding of strains or hybrids that are relatively immune to attack by specific pests.

Domestic animals grazing in forest areas, particularly cattle and goats, often cause much damage by killing seedlings and compacting the soil. In the eastern Mediterranean region, where goats have been allowed to roam over the countryside, much of the forest vegetation has been completely destroyed. In many cases the only control possible is to exclude the animals; elsewhere forest grazing may be permitted under carefully controlled conditions.

Forests are normally protected against the adverse effects of weather and exposure by some adjustment in the methods of cutting and by encouraging the development of species best suited to local conditions.

Silviculture

In essence, silviculture is applied forest ecology and forest genetics; it may be described as the development and use of cutting methods and cultural treatments of the forest:

- To ensure adequate regeneration of desirable species as soon as possible after the mature stand is cut,
- To bring about conditions favorable to the optimum yield and quality of production in keeping with the objectives of management and the condition of the forest
- To maintain and where possible improve the quality or productivity of the site.

Reproduction

A tree crop may be established by artificial or natural means, or by a combination of the two. In artificial reproduction, seed is sometimes sown, but more commonly young trees that have been grown in nurseries are planted on the site. In either case, the seed used is collected in the forest, preferably from trees of good form and vigour. When the crop is established by natural means, the young trees may originate vegetatively or from seed disseminated either before felling takes place or by trees left standing for the purpose at the time of logging. A new crop of young trees that becomes established in the forest before cutting takes place is known as "advance growth."

There are two principal methods of cutting the mature forest, each of which has advantages with respect to the regeneration and establishment of individual tree species:

Clear-Cutting

By this method, which results in the development of more or less even-aged stands, the forest is cut in one operation or in a succession of operations, each of which removes either a portion of the trees scattered over the whole area or all of the trees from a part of the area; these operations are continued until the original stand is completely cleared. Regeneration may be established by natural or artificial means, the former being provided for either by advance growth, or by seed lying in the surface soil at the time of cutting or made available from the crowns of felled trees or from standing trees surrounding the cutover area.

For species that require overhead protection in their youth a "shelterwood cutting" is made: the stand is opened up gradually over a period of years by making three or four successive cuts periodically over the whole area until the original stand is removed. For other species, the seed source may be the critical factor and the stand is clear-cut except for a few mature trees, evenly distributed over the area, which are preserved to provide the seed for the next crop. This is known as "seed-tree cutting." Quite commonly, portions of a stand are clear-cut periodically in the form of narrow strips, wedge-shaped blocks, or patches, until the stand has been completely felled.

The Douglas-fir forests on the Pacific coast of North America, for example, are clear-cut in patches or blocks of a size that will permit natural regeneration by seeding-in from the surrounding uncut stand.

This type of cutting is particularly applicable to species like the Douglas fir that are intolerant of shade. The intensity of cutting and the size, form and arrangement of clearcut areas vary according to local conditions and requirements, particularly stand density, topography, logging methods and the ecological characteristics of the desired species.

Selection Cutting

This method distributes cutting and regeneration continuously over the whole area, producing a single uneven-aged stand or series of such stands where trees of all ages from seedlings to mature timber occupy the area together. Scattered single trees or small groups of trees may be removed periodically throughout the whole forest, or the forest may be divided into a number of blocks, one of which is cut over in this way annually or every few years.

The aim is to improve and regenerate the forest, as well as to harvest a crop: while some of the better formed mature trees may be kept in the stand for a number of years to provide seed, all dying and defective trees that are at all merchantable are removed as quickly as possible. Nonmerchantable trees may be girdled. As regeneration and early development of growing stock must take place in partial shade, this method is particularly suited to tolerant species.

Afforestation

Forestry not only seeks to ensure the reproduction of existing forests, but is also concerned with the problem of establishing new forests. Afforestation may be undertaken *(a)* to reclaim forest land cleared for other purposes but best suited for the growing of forest crops, *(b)* to establish forests as an economic crop on land not previously forested, *(c)* to stabilize shifting sand and eroding soils, *(d)* to provide protection for agricultural and other crops against wind and snow damage and *(e)* to improve seepage in drainage areas and so stabilize stream flow.

Great Britain, Australia, New Zealand and South Africa have all undertaken large-scale afforestation with the object of supplementing their natural resources.

The trees used have been predominantly softwoods, particularly species of pine. One of the best-known examples of land reclamation for forestry purposes is in the Landes of Gascony, where, since the latter part of the eighteenth century, a programme has been developed to fix shifting sand dunes, to drain swamps and to afforest large areas with maritime pine the principal source of resin and turpentine in Europe.

The use of trees in shelter belts and windbreaks to protect agricultural land is common practice in many countries and has been practiced on a large scale in the United States and Russia. The establishment of forests to conserve water supplies has often been an important objective of forestry ever since its early beginning in Europe. In many cases, however, such forests may be used for wood production as well as for protection. In India and certain other tropical countries forests created primarily to provide small timber and fuel for the local population often help to solve the problems of flood control, erosion and desiccation.

The success of afforestation programs depends upon many factors. Foresters must determine the suitability of species to soil and climatic conditions; they must guard against the many hazards that may arise in introducing a species or race to a new environment; they must decide whether the species grows best in a pure stand or in mixture with other species and whether it will thrive as a pioneer under open-grown conditions or is better suited to a later stage in plant succession. The relationship of the species to its environment must be studied carefully if afforestation is to be successful.

"Tending" Operations for Yield and Quality

Once the forest is established, the aim of the silviculturist will be to develop and maintain the optimum growth and quality of the stand in keeping with the objectives of management. This is accomplished mainly by cutting operations of various kinds and intensities, especially by means of thinnings carried out periodically throughout the life of the stand. These are commonly referred to as "tending" operations, as distinguished from those aimed primarily at regeneration.

Although forest trees are members of a community and, as such, contribute something toward each other's well-being — e.g., mutual protection against the elements and enrichment of their common habitat, the forest soil — there is, as noted above, a continuous and sharp competition among them for light and space and in some cases for soil moisture and mineral nutrients. The stand density (number of stems per acre) greatly affects the intensity of this competition and also influences the rate of growth per tree and the ultimate value of the forest to man.

Crowded conditions and keen competition are usually beneficial in the very early life of the stand: the inherently vigorous individuals become dominant, the lower branches of the trees remain small and are killed back at an early age and the tree tends to develop a straight bole.Later, however, too much crowding is a disadvantage. With reduced incidence of sunlight, radial development of the crowns is restricted and growth is slowed down.

In fact, stagnation may take place, especially in species showing little differentiation in height growth. It is the aim of the silviculturist to make the most of early competition while this is advantageous and to remove it when it hinders stand development — in other words, to distribute the total productive capacity of the site among the optimum number of trees to obtain the desired results. This is usually accomplished by thinning operations.*Thinnings.* Cuttings made in a fully stocked to overstocked immature stand in order to increase the rate of growth of the remaining trees are known as "thinnings." In general, these are begun in the late sapling stage and are continued periodically until the stand matures: they are usually of only moderate intensity until rapid height growth is completed, the later thinnings being more severe to accelerate diameter growth.

The total annual growth of an area is not materially changed by thinning, but it is confined to fewer trees whose individual increments are thereby increased. As a result, the trees reach useful sizes at earlier ages, which is particularly important when logs of large size are desired for sawing into lumber. Too heavy a thinning, however, may leave insufficient trees for full use of the productive capacity of the site.The economic feasibility of thinning operations is largely determined by the availability of markets for small-sized material, although in some cases the improved growth of the residual trees may justify the operation.*Other Cultural Treatments.*

To promote more desirable species and better formed trees and generally to improve the quality of the stand, other cultural treatments may be used, among them cleanings, liberation cuttings and improvement cuttings.

- *Cleaning.* An operation in a young stand not past the sapling stage:
  - To free small trees from weeds, vines, or sod-forming grasses;
  - To provide better growing conditions by liberating crop trees from other individuals of similar age but of less desirable species or form that are overtopping or likely to overtop them.
- *Liberation Cutting.* The release of young trees not past the sapling stage from competition with older trees that are overtopping them.
- *Improvement Cutting.* A cutting made in a stand past the sapling stage to improve its composition and character by removing trees that are less desirable with respect to species, shape, or main crown canopy.

Pruning. Knots in lumber are the result of branch growth; the fewer and smaller the branches, the fewer and smaller the knots and the higher the grade of lumber produced. In a young stand of proper density the lower branches are often small in diameter and die at a fairly early age.

The forester may assist in the production of knot-free wood by removing the dead and dying branches from the bole of the tree with axe or knife. Care must be taken, however, not to injure the tree nor to reduce unduly the living portion of the crown. A comparison of the cost of pruning with the relative value of knot-free wood will determine the feasibility of this treatment.

Silvicultural Systems

As a result of past study and experience, particularly in northwestern Europe, there has been developed a series of silvicultural systems, each incorporating the techniques that are required in the silvicultural management of a specific type of forest to maintain its distinctive form and maximum productivity.

While these systems are based primarily on the method of regeneration (whether by coppice or seeding and with reference to the form of cutting), they are also concerned with problems affecting the protection and tending of the forest and the harvesting of the crop and its economic utilization. Where such systems have been evolved, they will form an essential element in the plan of forest management.

Forest Management

Although silviculture and management may be looked upon as two branches of forestry, the one mainly technical and the other economic, the two are interdependent and must always be coordinated. While silviculture may establish the ideal possibility of production within a forest, management determines the degree to which this can be realized through regulation and within the bounds of good business practice. Forest management may be defined as the application of business methods and technical forestry principles to the operation of a forest property.

Compared with other crops, a forest requires a relatively long time between its establishment and its harvesting. Revenue from an unmanaged forest may accrue only at fairly long intervals. Managed forests are organized to ensure a sustained yield of the forest crop in perpetuity. The income should include a reasonable profit on the investment, annually if possible.

Sustained yield of timber depends upon the systematic reproduction of a crop as it is harvested. It also requires that the timber be cut annually or periodically in such quantities that a continuous and fairly uniform yield will be provided throughout the rotation (the period required to grow a crop from seed to maturity). The principle of sustained-yield management may be illustrated with reference to a theoretical forest, usually termed the normal forest, in which the stands are stocked with the maximum amount of wood that the site will produce and from which the same amount of wood product may be harvested annually forever-namely, the amount that grows annually on the whole acreage. Consider, for example, an even-aged forest that is to be harvested when fifty years of age. To have fifty-year-old timber every year, there must be a series of fifty stands of, respectively, one, two, three and up to fifty years of age. Each year one stand becomes mature for harvesting. This stand in volume represents its own increment or growth for fifty years; also it is equal to the growth of one year on the fifty stands making up the whole forest. Accordingly, each cut is equal in amount to the current year's growth. It is assumed that each stand is reproduced as cut and that all the immature stands receive the silvicultural treatment necessary for most satisfactory growth. An alternative method, for an uneven-aged forest, is to cut over the entire area periodically, removing at each cut only the volume corresponding to the growth during the period between successive cuts. In either case, one has permanent sustained yield from a forest characterized by normal distribution of age classes, normal increment and normal growing stock.

There is, of course, no such forest. Every forest is abnormal, having either excess or deficiency of area in the older age classes and the opposite in younger ones and in general a deficiency, to a greater or less extent, in volume per acre in the stands of different ages.

The abnormalities in age-class distribution can be removed only by adjustments in the areas cut over in successive years, while deficiencies in volume, caused mainly by inadequate reproduction, can be made up through improvement in silvicultural practices. The normal forest is an ideal, a standard toward which forests should gradually be developed.

Such intensive management requires accurate information about tree species and sizes, total and merchantable volume, age-class distribution, growth conditions, site classes and rotation ages, as well as about such factors as topography and drainage, which affect the efficiency of operations. Such information is obtained through enumeration or inventory surveys, supplemented and improved in recent years by aerial photographs.

On the basis of this information, a management plan or working plan is evolved, which sets down the objectives of management and how they are to be attained and allocates the kind and amount of timber to be cut each year or period of years, at least for the first part of the rotation.

Logging

Logging engineering is concerned essentially with the felling and removal of timber from a forest area. Even where the sole purpose is liquidation of the resource, logging techniques may be developed to a high degree of efficiency, as they have a direct and important bearing on the cost of extraction. In forests under sustained-yield management, logging methods must be developed to meet the requirements of the management plan as well as to maintain operational efficiency.

Logging everywhere presents many technical problems, but particularly where the forests are difficult of access, distances are great and the topography is rugged. In many forest regions logs are transported long distances and at relatively low cost by natural water systems and in the northern forests winter snows often facilitate the hauling of forest products.

From the point of view of forest conservation, perhaps the most important logging objectives are:

- To prevent damage to the advance growth and the residual stand,
- To help prepare the site for future tree crops (e.g., by soil scarification),
- To ensure minimum waste of wood in the trees cut.

In recent years mechanical equipment has been increasingly employed in both the felling of trees and the transportation of logs.

Utilization

While it is not usual for those responsible for the growing and harvesting of timber crops to be directly concerned with their conversion and manufacture into usable products, there should be a close liaison between the two operations of forest production and wood utilization. Wood has inherent qualities that make it particularly suitable for many purposes—good strength in relation to weight, pleasing appearance, insulation against heat and sound, ease of fastening with nails, dowels, screws, or glue and case of working with relatively simple tools. However, wood is by no means a homogeneous material: it differs from one species to another in structure and physical properties and hence in the use for which it is best suited.

Within environmental limitations it is desirable to grow the species that will best meet the requirements of local industry and available export markets. Furthermore, to make the fullest use of the timber crop at all stages of development and to minimize logging and manufacturing waste, it is important to co-ordinate as closely as possible forest operations and the wood-using industries and also to bring about a closer integration of the wood-using industries themselves — as, for example, the sawmill and pulp industries.

Forest Policy

In many countries forests are of vital importance to the over-all economy, not only in relation to industrial development, but also in providing benefits which may be difficult to evaluate in monetary terms.

For this reason and also because of the long-term nature of the forest enterprise, most countries seek to ensure the perpetuation of their forests and to safeguard the interests of present and future generations by placing the forests under some degree of government control. In some countries the government retains ownership of a large proportion of the forested land and leases it to the wood-using industries for timber extraction: provision may be made for some degree of forest management by regulation or through agreement. Other countries have granted or sold much of their forest land to private interests, though considerable areas have sometimes been retained by the state to provide for timber reserves, recreational facilities and the protection of drainage areas; in some cases there is considerable public control of forest management on the freehold lands.

Elsewhere long-term forest management is entirely neglected and forest policies, if any exist, may be concerned only with the method and intensity of exploitation. Progress in the development and practice of forestry varies widely throughout the world. While forestry is well advanced in much of Europe and in certain tropical regions, only a small proportion of the accessible forests of the world is under intensive management today.

In some regions man still depends on the forests for his food and shelter and contributes little toward their development; in many countries man long ago destroyed the forests and has done little to restore them; elsewhere uncontrolled forest exploitation is still in evidence and proper protection and management lie in the future.

In North America we are still in an early stage of forest conservation. In recent years, however, we have begun to see the forests as an integral part of our economic structure and to take steps against the destructive exploitation that has threatened to exhaust them.

3

Ceylon and Forest Zones

Although Ceylon lies entirely south of the Tropic of Cancer, no place in the island is more than seventy miles from the ocean and the temperatures are correspondingly moderate. The monsoons (northeast from November to February and southwest from May to October) and a hilly relief combine to create diverse climatic conditions that give rise to an exceptionally rich variety of plant communities. Between Colombo on the west coast and Batticaloa, 140 miles away on the east coast, there are no fewer than six main climatic and corresponding forest zones.

THE TROPICAL EVERGREEN ZONE

The southwestern part of the island is covered by tropical evergreen rain forest similar to that usually found in warm tropical regions of heavy rainfall and high humidity. This region benefits from both monsoons and receives a well-distributed rainfall of over 100 inches per annum; the mean annual temperature is approximately 80° F., with small diurnal and monthly ranges.

The vegetation is luxuriant and trees usually attain a height of more than 100 feet; climbers abound. The forest growth consists of a large variety of species producing timber mainly of nondurable quality. Competition between species is so great that it is not easy to select any one as being typical of an area. In certain localities Dipterocarpus zeylanicus produces almost pure stands. Other strong competitors are Doona congestiflora (thiniya), Cyathocalyx zeylanicus (kekala), Calophyllum bracteatum (kina), Artocarpus nobilis (del), Cinnamomum multiflorum (wal-kurundu), Ficus spp., Melia composita (lunumidella), Palaquium grande (mihiriya) and Pericopsis mooniana (nedun).

THE WET MONTANE. ZONE

Northeast of the tropical evergreen forest zone lies the wet montane zone. A steady decrease in the luxuriance of the tree growth with altitude is observed. The species found at lower elevations begin to disappear and others take their place. The more prominent among the latter are Michelia champaca (sapu), Syzigium umbrosum (damba), Acronychia pedunculata (ankenda) and Kurrimia zeylanica (palen). Above about 4,500 feet, the forest becomes a true montane evergreen forest with species such as Syzigium umbrosum, Syzigium rotundifolium,Gordonia zeylanica, Meliosma simplicifolia (elbedda) and Calophyllum walkeri (kina). Higher still the trees are even more stunted and branched. Dead and dying trees are seen practically everywhere.

Although the annual rainfall is over 100 inches and is well distributed throughout the year, strong winds and very low temperatures from December to February (26.1°F. recorded in February, 1914) prevent good forest development. This type of forest alternates with undulating grasslands locally called patanas, on which the only tree species is the sparsely distributed Rhododendron arboreum (ma-ratmal). The shifting cultivators who originally cleared the land of its forest cover, together with the annual grass fires that followed, may be responsible for the almost total absence of forest in these areas.

THE INTERMEDIATE MONTANE ZONE

East of the wet montane zone is an area of lower rainfall (about 75 inches yearly), shallow soil of low storage capacity and droughts of four months or longer in duration-conditions not conducive to the development of trees. The montane evergreen forest of this intermediate zone is confined to watercourses, folds in the hills and other places with a moist soil and consists of a mixture of species found at both higher and lower altitudes. Below 3,000 feet occur species characteristic of the adjacent savanna forest of the lower elevations, such as Terminalia chebula (aralu), Terminalia belerica (bulu), Emblica officinalis (nelli), Anogeissus latifolia (dawu), Butea monosperma (gaskela) and Pterocarpus marsupium (gammalu).

Lantana aculeata, originally introduced into Ceylon as a flower plant, has, in many places, become a persistent weed. The large extent of grassland (dry patanas) that alternates with these patches of forest is due to the enormous runoff over steep slopes, to long periods of drought and to the shallowness of the soil above the gneiss rock, although in some localities shifting cultivation and subsequent grass fires have been responsible. The only tree species which survives in these dry patanas is Careya coccinea.

THE SAVANNA FOREST ZONE

At approximately 1,500 feet one enters the parklike savanna forest, locally called damana or, in the wet zone, talawa, an apparently unending carpet of grass with a few scattered trees. The annual rainfall here is more than 75 inches, but the period of drought in some years lasts as long as five months.

This country of indescribable beauty is still the haunt of the elephant, buffalo and leopard. A paradise for the hunter, landscape painter and poet, it is of no economic value as a forest. There is a total absence of undergrowth and the herb layer consists of grasses, of which Imperata cylindrica (illuk) is the most prominent. Shallow, poor soil, long periods of drought and the destructive activity of man have contributed to the making of this plant formation.

THE SEMI-EVERGREEN MONSOON DRY FOREST ZONE

This lies east of the savanna forest zone and has a similar climate. Most of the trees are deciduous and their average height is about 60 feet. These forests produce some of the best timbers in the world -- Chloroxylon swietenia (satinwood), Berrya cordifolia (Trincomalee wood, halmilla), Vitex pinnata (milla), Diospyros ebenum (ebony), Alseodaphne semecarpifolia (ranai) and Manilkara hexandra (palu). These valuable species grow in mixed stands with others less valuable, such as Hemicyclia sepiaria (wira), Ficus spp., Walsura piscidia (kiri-kon), Pleurostylia opposita (panaka) and Aglaia roxburghiana (kannakompu).

THE COASTAL JUNGLE ZONE

The arid coastal belt in the east, northwest and southeast of the island is covered by a thorny scrub jungle, which does not exceed 25 feet in height. This zone receives less than 50 inches of rain concentrated in about four months of the year. The air humidity ranges from 60 to 75 per cent. The plant community consists of shrubs such as Gmelina arborca (et-demata), Randia dumetorum (kukuruman) and Memecylon rostratum (kuratiya). The scattered tree species found here are Cassia fistula (ehela), Calophyllum calaba (gurukina), Manilkara hexandra (palu), Hemicyclia sepiaria (wira), Euphorbia antiquorum (daluk) and Chloroxylon swietenia (satinwood). Commercially these forests are of no value, but merely provide cover for the poor sandy soil.

Since climatic changes are not abrupt there are few sharp divisions in the forest vegetation: the zones merge into one another, forming intermediate communities. The area between the tropical evergreen and the semi-evergreen monsoon dry forest is covered by a forest that has some of the characteristics of both, but most of this plant formation still awaits proper scientific description.

DESTRUCTIVE FACTORS

SHIFTING CULTIVATION

The activities of the shifting cultivator have contributed much to the establishment of certain of the present plant communities. Owing to its good accumulation of humus adapted to crops, high forest is invariably cleared and burned and thousands of cubic feet of valuable timber go up in smoke every year. This method of cultivation, which ignores sound agricultural principles, is responsible for covering large areas with secondary forest (jungle) of no commercial value, or else with grasses. Accelerated soil erosion in such areas is enormous, especially on steep slopes. Tradition and lack of suitable land to satisfy all chena (shifting) cultivators have prevented the government from taking drastic steps. To localize shifting cultivation, however and thereby to prevent the wholesale devastation of valuable forests, the government has divided the chena land into blocks, each of which is cultivated once in ten or twelve years and has also made available irrigable land for permanent cultivation. Areas subject to heavy chenaing are now being reforested on a co-operative basis according to the "taungya forest system," which was first developed in Burma in 1856 and has been adopted in Java and other countries of Southeast Asia. Under this system government land is leased out rent-free for the raising of short-term food crops and in return the lessee raises approved forest species along with his crops.

COMMERCIAL PLANTATIONS

Clearing for commercial plantations has been the second most important destructive human factor. Since 1873 more than half a million acres have been opened up for tea plantations and since 1876 nearly 700,000 for rubber (Hevea brasiliensis). Most of these areas were previously covered by forest. Soil erosion on the plantations is heavy and has adversely affected rivers, tanks, irrigation channels and the plantations themselves.

ILLICIT FELLING

A third factor altering the composition of plant communities in most forest areas is illicit felling and the extraction of poles for agricultural and building purposes. Illicit felling is locally looked upon as an honest trade, one that has come down from father to son. Extraction of poles, seldom properly supervised, tends to destroy nearly all valuable timber species in the round-pole stage. These sizes are extensively used for fencing cultivated land and the construction of semipermanent buildings. Despite the fact that "the round pole of today is a log of tomorrow," removal of poles in other than silvicultural operations is still permitted.

Forest History and Administration

When Ceylon was ruled by the Sinhalese kings the entire forest estates of the island belonged to the Crown and large areas were cleared for agriculture based on irrigation. After the eighth century, however, the forest came back, owing to the destruction and neglect accompanying continuous wars with invaders from the continent and to the ensuing ravages of malaria. The forest at first reclaimed the Anuradhapura area, including the large city of the same name and, a few centuries later, the city of Polonnaruwa.

But in the later forest history of Ceylon "man the destroyer" features prominently. Export of timber began with the Portuguese occupation in the sixteenth century. The Dutch destroyed much forest for the cultivation of cinnamon, in which they carried on a lucrative trade and after 1830 the British undertook timber exploitation on an enormous scale. After several million cubic feet of the best timbers had been exported, a forest policy was initiated under British rule in 1885 and since then conservation and propagation of the forests have been promoted by the government.

The total forest cover is approximately 70 per cent of the island's area. Forest administration is at present the responsibility of four different bodies. The major portion of the forests is administered by the Forest Department, with a Conservator as head. The Department of Wild Life administers forests in the national reserves and sanctuaries for the flora and fauna. The Gal Oya Development Board, an autonomous body created for the multipurpose development of the Gal Oya valley, administers the forests in that area and the Revenue Department administers village forests set apart for domestic needs. Private forests form a negligible part of the total forested area.

Care of the Forests

Timber extraction for domestic and commercial purposes has been heavy in the past, but the inroads made during and since World War II have contributed most toward the present depletion of timber stocks. A considerable period of rest and cultural treatment is necessary for the majority of the forests, except where inaccessibility, lack of drinking water in the dry seasons and labour difficulties have prevented exploitation and thus helped to protect the stock. In the tropical evergreen zone the forests of the Sinharaja region are still intact. The stocks here, which consist mostly of Dipterocarpus zeylanicus (hora), could, on a rough estimate, produce half a million cubic feet yearly. In the dry-forest zone, the main source of valuable timbers, the only unworked forest is one of about half a million acres in the Northern Division (northwest of Trincomalee).

EXTRACTION AND REGENERATION

In Ceylon today modern machinery and methods applied in biblical times are found side by side. Owing to inadequate staff and consequent lack of supervision, timber contractors used to apply the "cut what you like" policy and the present generation is paying heavily for the sins of its forbears. Now, however, the extraction of timber is usually done on a selective basis and precautions are taken to leave sufficient trees of valuable species as seed bearers.

The heavy intermixture of species renders the removal of valuable timber, which seldom represents more than 5 to 10 per cent of the stock, both difficult and costly. The best forest tractor, one that can not only move easily between bushes and unwanted trees but also bring logs to the roadside and load them into trucks, is still the elephant and elephants are also employed in almost every sawmill yard.

Regeneration of valuable species after extraction is quite satisfactory, although species of low commercial value, such as Xylopia parvifolia (netaw), Ailanthus triphysa (wal-bilin) and Canarium zeylanicum (kekuna), in the wet zone and Ficus spp. and Hemicyclia sepiaria, in the dry zone, tend to overdevelop in areas that have been heavily exploited. Clear-cutting or excessive opening usually results in areas being overrun by forest weeds such as Macaranga peltata (kenda) and Trema orientalis (gedumba).

SILVICULTURAL OPERATIONS

Creepers and unwanted trees are removed by felling, girdling, or poisoning and thinning is confined to plantations. Plantations in the dry and intermediate zones were started in 890, as part of a plan for the co-operative reforestation of areas opened for shifting cultivation. About 15,000 acres of good forest have been planted. Species introduced with varying success are Artocarpus integrifolia (Oak), Swietenia macrophylla (mahogany), Tectona grandis (teak), Azadirachta indica (margosa), Vitex pinnata (milla), Felicium decipiens (pihimbiya), Alstonia macrophylla (havari-nuga) and Berrya cordifolia (halmilla).

In some areas the forest stock has been improved by planting Berrya cordifolia in small clearings scattered within the original forest. These operations have been fairly successful. The afforestation of the grasslands (patanas) of the central hills comprising the montane zone was begun in 1929 and approximately 5,000 acres have been successfully planted with exotics (Australian gums and some conifers). Although no previous tests were made of the suitability of the species to local conditions, the choice has mostly proved good. The species that have shown themselves most adaptable are Eucalyptus robusta, E. microcorys F. Muell. and E. saligna Smith.

After trials the planting of E. globulus,E. acmenioides Schauer and E. maculata was abandoned. Some areas in the montane zone have been planted with conifers, mostly Cupressus macrocarpa and trials have been carried out with Juniperus procera Hochst ex. Endl., Araucaria and Pinus spp. Acacia melanoxylon R. Br. has been successfully raised, but only on a small scale. Conifer planting was abandoned for some years owing to the extreme susceptibility of the plantations to damage by fire. A fresh experimental attempt, however, is being made with Pinus caribaea Morelet. Except where strong winds prevail, all plantations on up-country grasslands show good promise. With the cooperation of the FAO, the Forest Department has embarked on an extensive scheme of mechanized afforestation of the patanas and low forests and at present the target is a minimum of 2,000 acres per year.

Chief Timber Uses

In Ceylon timber is used not only for building construction, which makes heavy demands, but also in the manufacture of plywood, furniture, chests, safety matches, toys and railway sleepers. The demand for nondurable light hardwoods is on the increase. A plywood factory run by the government utilizes such species as Doona congestiflora (kokun), Calophyllum bracteatum (walu-kina), Palaquium petiolare (kirihembiliya) and Myristica dactyloides (malaboda). More than 150,000 cubic feet were used for this purpose in 1952. Species consumed in the manufacture of chests, toys and matches are all nondurable varieties such as Alstonia scholaris (ruk-attana), Canarium zeylanicum (kekuna) and Antiaris toxicaria (riti).

The furniture industry, on the other hand, uses the most valuable durable timbers and there is a yearly demand for approximately 200,000 sleepers (600,000 cubic feet), which can be made only from primary hardwoods. Wood is also extensively used as fuel in the processing of tea and rubber and for domestic purposes, but this requirement should eventually be met by the development of hydroelectric schemes, one of which (Laxapana) is already under way.

The timber position of the island as a whole is unsatisfactory, extraction at present exceeding the yearly increment. The population of Ceylon, now well over 7,000,000, is growing rapidly and at the present rate of increase will double itself in another 25 years. Concurrently the demand per capita for timber, now approximately 10 cubic feet, is also rising.

The problem of future supplies cannot be solved solely by afforestation and better forest management, which can never keep pace with the demand. During the last dozen years there has been an alarming increase in the import of timber, the value of which has jumped from Rs. 193,000 in 1942 to Rs. 21,000,000 in 1952. The only solution is the utilization of species that have hitherto been treated merely as forest weeds. The Gal Oya Development Board has taken the first step in this direction by installing a plant for the pressure treatment of timber and even secondary species (light hardwoods) are now used successfully for building purposes. The Forest Department has also started experiments in timber treatment in a small experimental pressure plant and in open tanks and steps are being taken to install a plant similar to that of the Gal Oya Development Board.

With a view to the utilization of waste timber material from sawmills and carpentry workshops and of branch wood and inferior species, efforts are now being made to develop building bricks from a mixture containing timber and cement and production on a small scale has already been begun.

Governmental and Public Interest

Of recent years both government and the public have shown an increasing awareness of the value of forests as a national asset and of the important role they play in the country's economy. To raise the standards of forestry a forestry field-training school was established in 1951 with two training centers, one in the wet and the other in the dry zone and considerable research work is carried on by the Forest Department. The Soil Conservation Act, passed in 1951, empowers the Government to enforce anti-soil-erosion measures even on private lands. All these steps augur well and it is expected that in the not too distant future the present unsatisfactory timber position will be remedied.

There are three national parks, two of which-Ruhunu and Wilpattuhave been in existence since 1937. These abound in the fauna of the island and provide excellent relaxation and recreation for young and old. Ruhunu in particular, which is only 176 miles from the port of Colombo, has earned popularity in many parts of the world and even casual visitors to the island find time to spend a few hours here.

The third and most picturesque park, recently created under the Gal Oya scheme, shows promise of great popularity. The island has some of the most beautiful scenery in the world and forest work is made especially interesting when the forester finds deep in the forest, as he often does, ancient ruined temples and other monuments to a splendid past.

OCEANIA

Ten thousand islands -- thirty million square miles of sea -- less than 2 per cent of the surface land. One island alone, New Guinea, has three times the land area of all the others. Those others are as if an area equal to that of Great Britain had been fragmented and scattered over one third the world. Such is Oceania. Yet its islands and their forests, specked on some of the earth's most critical airways and seaways, of strategic importance in peace and war, command attention. The origins and relationships of the floras of Oceania have challenged the ingenuity, imagination and fact-finding capacities of botanists.

A drop of only 60 feet from present sea level would join New Guinea and Australia. Such a united continent of Papualand is believed to have existed 30,000 years or less ago. As a result of this or earlier unions, the forest flora of New Guinea bears a marked Australian character, especially as to casuarina, eucalyptus and oak. A lowering of only 60 or 70 feet more would expose the Sunda Shelf, uniting Southeast Asia with Sumatra, Java and Borneo. Such a union in the past must have brought the continental flora far east before its overseas migrations were launched.

THE COMING OF THE PLANTS

Although the last word has by no means been said, it would appear that the lesser Pacific islands obtained their original plants from across the seas, with all the hazards and uncertainties and failures of a man seeking to cross in a rowboat and trying to perpetuate himself if he could land. Hurricanes, rafts of floating logs, stray birds and other "accidents," continuing to this day, have contributed to the origins of the present floras.

There is now a consensus that the floristic invasion of Oceania has come from three main regions: the Americas, most obviously in Hawaii; a now vanished Antarctic continent, with the remnants of its flora found in southern South America, South Africa, Australia and New Zealand; and Indo-Malaysia, which in turn is basically Asiatic.

Despite the prevailing easterly winds, however, the tropical cyclones from the west have been most influential. As one proceeds from the western borders of the Pacific, the flora becomes more and more impoverished in genera and appears obviously to be a fraying-out, an attenuation of the Asiatic plant world, as the chances of successful immigration become less and less over the tremendous distances of water.

Fosberg has analyzed the Hawaiian flora with regard to its origins, basing his discussion on the number of assumed original immigrants. For the seed plants, he finds that 40.1 per cent are Indo-Malaysian, 18.3 per cent American, 16.5 per cent Antarctic and the remainder boreal, cosmopolitan, or obscure.Offsetting in part this paucity in genera and larger categories, there has been an extraordinary speciation in certain groups.

Whatever be the causes, the original immigrants, including trees, have sometimes indulged in endless morphologic variation, often with each valley developing its recognizable form.

Certainly many of the named forms are ecologic modifications and many more would interbreed and lose their identity if placed in a related population; but many others are undoubtedly established genetically and the recognition of them as desirable strains in forest trees is an open field for silvicultural research. Immigration and speciation are probably continuing now as in the past, but on too slow a time scale to be of practical importance. The striking effects of introductions by man will be considered later.

The above discussion holds for what one might term the "inland flora," for species that cannot tolerate salt, die when submerged in salt water and shun the strand. It is these species which pose the problem of overseas migration to the botanists.

The "strand flora," entirely to the contrary, seems to have evolved capable of overseas migration. Buoyant seeds that withstand many days of floating -- even as germinated seedlings -- are found in totally unrelated plant families. This flora is remarkably constant in all the islands -- evidence of the success of these adaptations; and this flora is adapted not only to migration but also to growth under saline conditions, either in soil or air. Indeed, the plants are frequently incapable of terrestrial migration (though many grow well where planted).

For all these reasons, the strand flora forms the entire vegetation on countless low islands. On the high islands it tends to predominate only in the plant communities near the seacoast.

Among all these native plants, as well as among naturalized species, there are innumerable kinds of trees of distinct value to the forester, especially in the islands close to Southeast Asia. No manual of dendrology exists for this territory, although Kraemer has published a volume for part of Malaysia, including some of Melanesia. The book describes more than 100 important species, said to be about one per cent of the total tree flora.

THE EFFECTS OF MAN

Black Men and Brown Men. The first peoples to be pushed out of Asia into the islands, either by deterioration of climate during the last glacial period or by more aggressive tribes, or both, were Negroids, Ainoids and Veddoids. As they entered Melanesia ("black islands"), the black fuzzy-haired Negroid element was predominant. Hunters and food-gatherers of an Old Stone Age culture, they probably had no marked effect on the forests of their islands.

Four to five thousand years ago, the New Stone Age culture spread to the Negroid islands of Oceania, a culture that turned the people into farmers practicing a destructive shifting cultivation that has persisted to this day and is of great significance from the standpoint of world forestry. To these folk the forest was and still is a hindrance. They cut and burn the trees and plant crops for a few years, until soil fertility presumably lessens. Then the site is abandoned and a new forest is cut. In the meantime the old site grows up to secondary forest. In a decade this is recleared and reused, but for a shorter period. Fire is omnipresent. Rhizomatous grasses invade, which successive fires only encourage by burning the tree seedlings.

The grassland, encroaching on the surrounding forest even without agriculture, is unmanageable and the farmer moves away permanently to new forest sites. New Guinea today is a patchwork of forest and grassland wherever (which is almost everywhere) a "burn-dry season" occurs, from sea level to the frost regions above the mossy forest. Forest remains only in swamps, along stream sides and where the natives have not yet entered, as in a no-man's land between warring tribes.

When such tribes annihilate each other, the forest replaces the grassland, to be destroyed by some future tribe. Since the roving agriculture is a one-way process, especially with increasing population, this destruction of much timber of high commercial value is a problem of first importance to forestry. In addition, it poses a more general problem: the Melanesian in effect destroys his own habitat, for the grassland is useless to him.

After the New Stone Age had spread to Melanesia, but still before the Christian Era, paler, more Mongoloid people invaded the islands to the north, the area now known as Micronesia ("tiny islands"). And then the far-flung islands from Hawaii to Samoa and New Zealand were populated by the handsome race known as Polynesians ("of many islands"), who conquered vast distances in frail canoes. Ethnologists now consider these folk to have migrated through Melanesia toward the east and to have emerged with their present bronze skin.

The theory that they came from Peru, by the Kon-Tiki route is not fashionable. All these invasions were barely completed by A.D. 1400, the date of the first entering of Hawaii.

The Micronesians and Polynesians radically altered the forests of the high islands with regard to both flora and vegetation. They brought with them a limited number of plants; these included a few weeds, but for the most part were food plants, of which the coconut has done the most to change the landscape. It is now dominant on practically every low island and at the lowest elevations of the high islands too. What the forest was originally, or would be now if left to itself or planted to other trees, is a matter of conjecture and future experimentation.

A shifting upland cultivation, such as that of Melanesia, was not part of the culture on the high volcanic islands and consequently the mountain forests were but seldom entered. On the other hand, fire -- as always with primitive peoples -- was uncontrolled and often started for pleasure and hunting. It is the author's opinion that on many islands fires burned as frequently and as far up the hills as the vegetation and the climate would permit, destroying not only large areas of original forest but also the evidence of present forest potentialities.

White Men and Yellow Men. In 1520 Magellan passed through the straits that bear his name and, the following year, landed at the Mariannas. Thus was opened a new chapter in the exploitation of Oceania, a chapter of conflict that continues today between two fundamentally opposed philosophies of social living, a conflict that every forester faces squarely when he tries to administer tribal forest lands.

The original natives lived in small social units. Theirs was an agricultural and subsistence-barter economy, based on few material wants. With the crudest of tools, they fashioned a satisfying existence on infertile specks of land. Living under a form of "communalism," the individual had no incentive to amass personal wealth, for he would only have to share it with his tribe. To this day and for these reasons it is difficult to lure natives to forest work, unless they make a total break with their home villages.

In striking contrast, the new invaders were members of large social units. Theirs was a complex money economy, based on many and evergrowing material "necessities" Under the guise of "individual liberty," each man had every incentive for personal gain and thus wave after wave of Occidentals has passed over Oceania, each intent on its own brand of personal gain: souls, morals and modesty; loot and trade. The invaders streamed in from Europe, the United States and Japan and, when native muscle was not co-operative, Hindus, Chinese, Malayans, Filipinos, Javanese, Tonkinese and Puerto Ricans were herded in. World War II, however, left Oceania with a new political geography and under a "trusteeship" philosophy that has shown itself more altruistic and more humanitarian than history has usually known. Such elements make the Oceania of today the unusual stage that the forester must know if he is to act effectively.

Sandalwood, 1790-1850, is a finished case history of forest "mining" Undoubtedly, the value of island sandalwood was discovered independently on several occasions and the period of its exploitation is filled with double-crossings. Many a chief succumbed to the money lure, assuming island monopolies and forcing his people to work at gathering the wood in the unhealthy rain forests. Either by nature or with the connivance of the gatherers, who are said to have pulled up all the seedlings they saw, commercial sandalwood disappeared. Apparently the plant has some silvicultural peculiarities, for it has not reappeared in any quantity, either naturally or by artificial means.

The introduction of large naturalized mammals is another aspect of the Occidental invasion that is having a continuous and most important effect on the forests. Only the continental islands possessed a native fauna of large animals; the oceanic islands apparently developed their biotic communities in the complete absence of hoofed animals. It appears that the native forests have extremely shallow root systems, critically located in a moist surface layer. The slightest disturbance to the ground vegetation, as by grazing or browsing, kills the trees, in turn opening the adjacent forest to drying and destroying it, with an effect almost of a chain reaction. Cases are on record in Hawaii where grazing stock was put into a forest for a matter of a few weeks and then removed, but the forest died thereafter. If such destruction were on land planned for grazing, no unexpected problem would result.

Unfortunately the early white settlers introduced cattle, sheep, goats and horses. The islanders placed a taboo upon them, but some escaped to the hills and showed remarkable fertility. In addition, missionaries and traders often released animals on small uninhabited islands, for the sustenance of shipwrecked individuals. The destruction of the native vegetation by these feral animals and by pigs, earlier naturalized, has been tremendous. Wholesale slaughter of thousands of animals has since taken place in some islands; and to this day one of the most useful activities of a local forest service is the building and upkeep of border fences, not only to keep out adjacent high-quality stock, but to prevent reinvasion by feral animals.

Although rabbits have in general been less of a problem, Christophersen has made an interesting report on their history in Laysan, in the Hawaiian Islands. Until 1903 this little island had a flora of 28 species and a solid vegetation of grass and shrubs. In 1911 rabbits were swarming "over the island by thousands" By 1923 only four species of plants could be found and Laysan was described as a "desert of sand," with no mention of rabbits, alive or dead-another example of an introduced mammal annihilating itself by destroying its own habitat.

Much more than half of the flora of some of the islands is composed of species that have been introduced and naturalized since the white invasion. Moist uplands have been less affected, but arid lowlands -- with their original vegetation already destroyed-were fair prizes for hordes of weeds. In many islands the bulk of the lowland vegetation is now composed of alien species and one must search carefully for the indigenes.

Furthermore, on destruction of the native forest by grazing, the land is taken over almost at once by alien grasses and shrubs which, as on the grasslands of New Guinea, form a strong deterrent to the return of the native trees. This situation has given rise to egregious comparisons between the native flora and the native cultures that disintegrate before the onslaught of aggressive aliens. The contrary, however, would seem to prevail in the case of the flora. The writer (1942), in accord with Allen's interpretation for New Zealand (1936), has presented field evidence in support of the thesis that most of the alien plants are pioneer species expanding into "open" lands and forests that have already been disturbed, whereas the native trees are largely of end-stage (climax) species that will hold their own against most of the invaders. The critical need in reforestation is for a "nurse tree," native or alien, which can naturally invade the grassland and scrub, forming a growth that can in turn be invaded by the native forest trees.

Cimates

Oceania comprises three main kinds of islands:

- "Oceanic coral low islands," composed entirely of coral limestone and presumably capping submerged volcanic deposits;
- "Oceanic volcanic high islands," composed of active or extinct volcanoes, arising from the floor of the ocean basin and usually complicated by fringing reefs;
- "Continental high islands," composed of sedimentary strata overlying a granite base, usually complicated by volcanoes and coral reefs.

The low islands partake of what might be called regional climates, controlled by the warm moist trade winds blowing diagonally from the horse latitudes (at about 30° N and S) westward toward the equator. These winds, as they pass over warm, low, sandy islands, may be sufficiently heated so that no rain falls, even when there is precipitation on the surrounding ocean. In general, within five degrees of the equator there is a dry zone, which includes the Phoenix and Gilbert islands and at higher latitudes a moister zone.

At still higher latitudes, the seasonal shifting of the trade-wind belt with the prevailing westerlies modifies the climate, as in Hawaii. In the western Pacific, especially in New Guinea, the influence is predominantly that of monsoons, related to the greater seasonal warming and cooling of large land masses than of the adjacent ocean.

The climates of high islands may vary markedly even within distances of less than a kilometre. The moist trade winds, striking a volcanic island, are deflected upward, where they cool to condensation. Passing over and beyond the summit, they drop, are warmed and dry up. The result is a series of concentrically arranged precipitation belts, placed askew on the contour pattern.

The centre of highest precipitation is slightly to windward of the peak and each rainfall belt is tipped down to windward so that at any one elevation there is less rainfall and a drier vegetation on the lee than on the windward side of the island. Each valley and ridge adds its minor complications. The majority of these climates exhibit a great variety of physiognomic vegetation types, depending largely on human activities, especially management practices. Consequently, any climatic classification which invokes a single vegetation type as the "climax" for its zone runs immediately into inconsistencies.

Soils

Few generalizations can yet be drawn on the soils of Oceania. Those of the continental islands are probably as varied and as complex as in any continental area, though they are today little known.

The low limestone islands have invariably a thin and poor soil, occasionally favored by a layer of organic matter from the vegetation or from bird colonies. More has been published on the soils of the high volcanic islands, especially as regards agricultural Hawaii. It is probable that many of the forest soils are ferruginous laterites. Aside from a surface crust, sometimes present, aluminum and silicon increase downward, iron concentrates at an intermediate position and titanium remains abundant at the surface.

Soil erosion can be critical, often spectacular, but as yet has failed to arouse the governmental authorities, as in other parts of the world. Perhaps this is because most of the erosion has been in unpopulated areas from the start and has not resulted in the displacement of contemporary populations and agricultures. Much of it apparently was consummated before the advent of the white men, or else immediately after-owing to the naturalization of grazing mammals -- but before European land managers began to look more critically at the land. Within a few miles of Honolulu there was a valley forest a few years ago half buried with debris from the hills. Elsewhere entire coastal plains are said to have been formed in recent years and relict soil patches hanging on rocky mountains attest to past conditions.

Unlike that in temperate regions, the decomposition of the volcanic rocks in the tropics may be rapid and some lands may be amenable to management practices that would restore seminatural forests to areas now completely barren. The research on this remains to be done.

Island Types

It should be noted, however, that these are not only incomplete, but are not indicative of potential forest land, since there is much grassland and savanna that could be converted to forest under rational management practices. Such grasslands include not only those of New Guinea and the Solomons, already mentioned, but also those of Hawaii, Guam and even Yapy and Palau.

OCEANIC LOW ISLANDS

Where volcanic or other materials have formed the sea floor at depths of less than 150 feet in tropical waters, they have been subject to colonization by coral and related organisms. These have built up calcareous deposits as living reefs at, or slightly below, sea level. Surmounting such a reef, generally at its margins, an atoll is usually a more or less circular group of small islands, produced by wave and wind action, enclosing a "lagoon" To the geomorphologist atoll islands seem ephemeral, subject to rapid destruction, especially during storms. Forestry practice must reckon with the hazards to which such islands are subject.

Table. Oceania: Land Area and Forest Area for Certain Islands, 1950-1953 (In Thousand Hectares)

	Land Area	*Forest Area Accessible*	*Inaccessible*	*Total*
British Solomon Islands, 1953	2,890	20	2640	2,660
Fiji, 1953	1,828	140	815	955
French Oceania, 1950	360	50	65	115
Hawaii, 1950	1,665	450	325	775
New Caledonia, 1952	1,582	10	70	80
New Guinea (Australian) and Papua, 1951	47,400	3,00	30,000	33,000

As a habitat for plants, atoll islands, despite their almost infinite number, their human populations and their emotional appeal, do not hold much for the forester interested in quantity. Their vegetation normally comprises a series of belts from sea to lagoon. Back of the sea beach is a scrub 2 to 5 meters high. Next is a narrow forest zone of Messerschmidia argentea and Pandanus spp. In the Marshalls and Carolines Terminalia litoralis is found in this Messerschmidia zone.

The greater part of the typical atoll is a forest which today is almost entirely composed of Cocos nucifera, topped sometimes by huge breadfruit trees (Artocarpus altilis) where the rainfall is heavier. The timber of breadfruit is excellent in almost every respect, but the tree is seldom cut since it serves primarily for food. It would appear that the original forest vegetation was composed of such trees as Pisonia grandis, Pandanus spp., Ochrosia parviflora, Pipturus argenteus, Hibiscus tiliaceus, Messerschmidia argentea, CaloPhyllum inophyllum, Barringtonia asiatica, Eugenia spp. and others. Pisonia may be the sole dominant on dry islands, such as Vostok. Some of the isolated eastern atolls may have floras composed of as few as five species, or even less, seemingly the result of chance migration.

Along the lagoon beach of most islands is a narrow strip of scattered trees such as Cordia subcordata, Hernandia ovigera, Terminalia catappa, Barringtonia asiatica, Thespesia populnea and, in rocky places, Pemphis acidula. Mangrove species, where found at all, occur in small depressions in the centers of the islands.

OCEANIC HIGH ISLANDS

The oceanic high islands display endless variations, but basically they consist of volcanoes. They not only occur in chains and groups but also in such total isolation as that of Easter Island. The original formations were pahoehoe and aa lavas, with bombs, ash, cinders, dust and other ejecta. Erosion may have advanced to such a degree as to leave knife-edge ridges and deep valleys (sometimes amphitheater-headed, because of greater precipitation upstream), footed by coastal plains.

Ancient coral reefs, developed at previous higher sea levels, may appear along an island's flanks. Soil maturation has progressed variably from the different parent materials, with the development of distinctive profiles. Despite this profusion of types, a pattern emerges of interest to the forester.

This zones, arranged consecutively from sea level to summit. Each zone, except in the driest, coldest and cloudiest sites, has a characteristic assemblage of forest, savanna and grassland, with certain apparent correlations with contributing factors, although implications of causality are still premature. Sunlight attains a maximum at sea level on the lee side of the typical island and decreases upward, with a marked diminution at the beginning of the socalled "cloud zone," a minimum in the mossy forest and then a rapid increase to a second, high-altitude maximum.

Precipitation varies inversely with sunlight, being low at sea level, maximal in the cloud zone and again low at very high elevations.

Cloudiness, although it may occur at any elevation, is of significance only within the cloud zone, where its intensity increases rapidly to a maximum at its upper level. Altitude, perhaps because of the ease of obtaining data regarding it, is mentioned ad extremum by botanists and foresters, but does not seem to exert any direct effects, at least upon the details of vegetation.

The distribution of floras in relation to vegetation zones has been considered only for Oahu, Hawaii. Despite its significance in forest management, both in control of natural stands and in planting programs, foresters have almost totally neglected this subject. Its importance is illustrated by the fact that in any one zone the predominant trees may be either (a) those that are there capable of maximum individual development and would not thrive in an adjacent zone, or (b) those that have been crowded out of an adjacent zone by competition with others and would actually develop more luxuriantly if planted in the adjacent zone with the competition lessened.

As a working hypothesis for Pacific high islands, areas characterized by six different floras:

- The strand flora, already discussed;
- A desert flora, which may exist on some islands where there is less than 35 inches of precipitation a year (some widely naturalized Opuntias may occur in this area);
- An arid-region flora on islands having roughly less than 50 inches of rain, with a maximum floristic complexity at the higher precipitation levels and attenuating out below to mingle with the desert plants (tremendous numbers of alien plants have been naturalized in this floristic area);
- A humid-region flora occupying a region of rain forests, with high percentages of endemic species and few naturalized aliens (many of the plants pass upward into the cloud zone, where present);
- A cool-tropical cloud-region flora;
- A flora or floras above the frost line. On some islands, above the cloud zone but below the frost line, the humid-region and arid-region floras are repeated, but in reverse order, as on the island of Hawaii.

The significance of an effective frost line on the islands of Oceania remains to be investigated. One might expect, judging from continental areas, that it would form a fairly distinct separation of tropical from nontropical conditions. The situation in New Caledonia and in the New Hebrides, Society, Bismarck and Solomon islands is unknown.

In oceanic Hawaii the flora at high altitudes appears to be nothing more than an extremely impoverished attenuation of plants primarily tropical, with almost no temperate American origins-mute testimony of the failure of aerial migration from that direction.

Continental New Guinea, on the other hand, has an extensive super-frost-line flora, but its geographical affinities are poorly known. All the super-frost-line regions hold great potentialities for the introduction and naturalization of temperate forest species.

Each vegetation zone possesses a complex of topographic variations, ridges, valleys and plains that make difficult the recognition of zonal differences. In those cases where a so-called rain-forest zone is predominant on coastal plains and a "montane-forest zone" on foothills and slopes, one has reason to wonder whether a climatically significant zonal difference actually exists.

The precise mapping of the boundary between two distinct zones has been attempted only on Oahu. Here the line was found to pass through a wide physiognomic variety of plant communities and to vary in altitudinal range by almost 1,000 feet, bulging seaward on the ridges, where the cooler temperatures may have offset the lower rainfall to permit an equivalent moisture effectiveness.

CONTINENTAL ISLANDS

The continental islands of Oceania include Fiji, New Caledonia, the New Hebrides, the Solomons, the Bismarck Archipelago and New Guinea. In them are found not only all the varieties of forest characteristic of coral low islands and oceanic volcanic islands, but also their own complexities, due to larger floras, more varied rock strata and topographic features and climatic regimes ranging from slight to complete dominance by the monsoons. Nevertheless, the same basic pattern of arid, humid, cloud and super-cloud zones occurs, if not in the orderly arrangement found on a single high volcano in the tradewind belt.

VEGETATION OF THE FOREST ZONES

MANGROVES

Mangrove vegetation is the constant associate of the shallow ocean shores, mudflats and river embayments of almost all of the high islands throughout Oceania. Only in Hawaii -- most distant from sources of dispersal -- were mangroves absent until relatively recent times. Yet they have been introduced here and are now developing semi-natural stands. Mangrove sites are rare on low islands.

As one progresses from open sea water landward to upland and also to brackish water, gradual changes occur in the vegetation, which is generally segregated in definite belts. The forest fronting the ocean -- usually much the largest in acreage -- is dominated by Rhizophora spp. Frequently just back of the front Bruguiera spp. are common. Farther landward are large numbers of other trees: Xylocarpus (carapa) spp., Sonneratia caseolaris, Heritiera littoralis, Avicennia marina, Lumnitzera littorea, Excoccaria agallocha and the colonial 2-meter-tall fern, Acrostichum aureum. The important thatch palm “nipa” (Nipa fruticans) is predominant in brackish waters.

Economic utilization of this vegetation is most highly developed in Fiji, where 52 square miles (33,000 acres), owned by the Crown, are commercially worked. On the basis of a mean annual increment of 130 to 150 cubic feet per acre, the lands are handled for fuel wood on a 40-year rotation, with Bruguiera and Rhizophora supplying 95 per cent of the take.

THE XEROTROPICAL ZONE

Dry Deserts. No survey has been made of the islands to locate those areas within the xerotropical zone where the precipitation is so light (certainly no

More than 15 inches) that they have either a distinctive desert flora, or a permanently nonforest vegetation due to other causes than human interference, or both. A few small areas in New Guinea are reportedly in this category and also a few lowlands in the Hawaiian islands, as well as some of the highest summits above the cloud zone but below the frost line. At these high elevations in Hawaii, however, where there are scattered dwarf outliers of the Sophora chrysophylla forest, the vegetation of pre-goat days has not been satisfactorily interpreted.

Some low coral islands, such as Jarvis, may likewise belong in this class. In these cases, however, the few plants present are those of the cosmopolitan strand flora and no distinctive desert vegetation is apparent.

It is important to distinguish between the dry deserts, which are correlated with the "natural" or nonhuman environment and certain large areas that now resemble such deserts because of the direct and indirect effects of man. The former are relatively unalterable, but the latter, at least in some instances, could be converted into forests and even into productive watersheds under long-range management.

Dry Forest, Scrub and Savanna. A zone of dry forest, scrub and savanna is generally found where there is less than 50 to 60 inches of rain per year. On typical oceanic volcanic islands it occurs at low elevations, especially on the lee sides. On very high volcanic islands it may again appear above the mossy forest, although such occurrences have not been adequately described. On continental islands it may occur at any elevation where the precipitation is low.

Very few detailed accounts of this vegetation exist, for understandable reasons. In its original form it has been totally destroyed by the natives and hordes of recent introductions have filled the vacuum. Since such areas seemingly bear only worthless herbs and shrubs, grazing rather than forestry has been their hoped-for use. Nevertheless, in all foot slopes in this zone in the Hawaiian islands the small tree Leucaena glauca can invade and replace various shrubs such as the ubiquitous Acacia farnesiana and, in turn, is locally invaded by still other tree species. Coastal plains have, within the past half century, developed semi-natural forests of Prosopis chilensis, 40 feet high, that are of economic value not only for firewood but for cattle shade, cattle forage (from fruits) and apiculture. It may well be that other valuable trees could be developed in this zone.

In New Guinea areas apparently comparable bear "savanna forest" The grasses are typically Imperata cylindrica (arundinacea) and Saccharum spontaneum. The scattered trees, growing to a maximum height of 50 feet, consist, at least in Papua, of several species of eucalyptus, including E. alba, E. papuana, E. clavigera and E. tereticornis, while elsewhere Melaleuca spp. and Casuarina spp. are common. As a source of fuel wood and post wood, these lands have value for both aborigines and European settlers. What other trees they might grow under fire management is totally unknown, for no relict unburned area has been described.

THE PLUVIOTROPICAL ZONE

This zone of pluviotropical vegetation, areally predominant on the rainy Solomons and New Hebrides and occurring on the intermediate slopes of almost all the high islands, continental and oceanic, includes what is popularly known as "jungle" (although that Hindu term in its original sense referred to secondary thickets on uncultivated land, perhaps even in arid zones). It embraces not only the rain forests, that profuse expression of tropical luxuriance made famous by Tarzan, but also much land now covered with savanna or with thickets of worthless shrubs.

In New Guinea burned savannas mixed with secondary forest extend completely through this zone to the lower limit of the cloud zone at 5,500 feet, with Casuarina nodiflora common at the higher elevations. Such a mosaic of savanna and forest has much in common with the Parang vegetation of the Philippines, where the same species of Imperata and Saccharurn are also dominant in the grassland phase. In New Guinea Octomeles sumatrana, an aggressive, valuable, 150-foot-tall floodplain tree, forms one of the few pure forest types in this part of the tropics. Nevertheless, it is all but impossible to pick any typical genera of the mixed forest. Castanopsis junghuhnii has attracted attention because of its frequency and its habit of occurring in pure patches of an acre or more. A few dipterocarps are present, but they never predominate as they do in the Philippines and Southeast Asia.

In New Caledonia savannas similar to those of New Guinea are widespread, but complex rain forests predominate throughout the island. Melaleuca leucadendron is a typical tree. Related species of Araucaria and Agathis occur at higher elevations.

Fiji has its talasinga (sun-burnt lands), rolling plains extending up into the mountains to altitudes of 1,500 feet. The scattered woody plants include Acacia richii, Casuarina equisetifolia, Dodonaea viscosa and Psidium guayava. The rain forests of these islands, from sea level to 3,000 feet, are complex floristically, but much less so than those of New Guinea. In Tahiti most of the vegetation is of the pluviotropical type and is closely related to that of other Polynesian high islands. Stenolobium stans, of Atlantic tropical America, is said to be on the march here, with armies of young trees ascending these slopes"

In Hawaii the forests of the lowest portions of this zone were destroyed either before white men came or within their memory. Vast acreages are now covered with lantana, guava and scrub that is despised by beast and man, by taxonomist and ranger.

And yet in one area of this zone the tree Psidium cattleianum var. lucidum is moving in to replace these scrub species and in turn acquiring an understory of native forest trees, as economically important a vegetational development as any observed by the author. In many areas the scrambling fern Dicranopteris linearis is especially aggressive following disturbance, blanketing the land and even small trees, so that some have been misled into considering it a climax.

Farther up the slopes occurs Acacia koa (Hawaiian mahogany), which by its growth form, height and the quality of its wood is the only important native timber tree, with a relation to past fires and with silvical characteristics that are not yet well known. The typical rain forest of Hawaii is predominantly composed of Metrosideros collina subsp. polymorpha, a tree which generally does not exceed 50 feet in height and 12 inches in diameter. Associated with it are a large number of endemic species, several of which have excellent potentialities, at least for watershed cover. On the gentle slopes of the larger islands, this same zone supports a range industry which is third only to sugar and pineapple in the islands' economy.

THE CLOUD ZONE

The vegetation of this zone is sometimes, but not always, separated into two relatively pronounced forest belts, especially in the continental islands.

New Guinea has a cloud zone with a markedly distinct flora of highaltitude continental origins. The lower belt ("mid-mountain forests") begins suddenly at the lower edge of the clouds, which may be as low as 2,000 feet, or as high as 6,000. It is the region with timber of highest commercial value, though fires set by the aborigines are destroying much of it, converting large tracts into savanna. Araucaria cunninghamii is one of the most striking, abundant and valuable of the trees. In addition, there are many oaks with Nothofagus just above. Eugenia, Cryptocarya, Sideroxylon and Podocarpus are among other trees of this belt.

The mossy forest, which in New Guinea begins between 5,500 and 8,000 feet, has one of the most distinctive appearances of all the world's vegetation. Here the trees seldom exceed 25 feet in height. Mosses and lichens cover and hang from everything, bloating the smallest twigs to ten or twenty times their diameter. The air is cloud-saturated, the moss water-saturated.

In New Guinea, Lane-Poole noted Eugenia and Podocarpus as most abundant. At one time, he observed a morning drought, when all the moss was brittle and exploded to dust at the touch, a condition that may help to explain the large areas of burnt mossy forest he had seen. Thus, even here native fires have their influence.

On Fiji the mossy forest begins at about 3,000 feet. In Micronesia it is found only on Ponape and Kusaie, beginning at about 1,350 and 1,450 feet respectively; here the most abundant woody plants are Alsophila ponapeana, Gynotroches axillaris, Syzygium carolinense and Astronidium ponapense. Samoa has a comparable type that begins at about 1,500 feet. In the Hawaiian Islands this belt, sometimes suffering from pig destruction, may start as low as 1,800 feet. The predominant woody plants, rarely different from those of the forest below, include Metrosideros, Bobea, Straussia, Fagara and Labordia.

On Maui and Hawaii islands, both the cloud zone and the forest just below are squeezed out of existence on the dry, leeward side and the dry forest that would otherwise lie below them ascends to higher elevations.

From the foregoing it can be seen that the mossy forest is highly variable in both altitude and floristic composition. Some of the variation in floristic composition is but a part of the floristic individuality of each island. The evidence is not yet conclusive as to whether the cloud zone tends to possess its own distinctive flora, or is a reduction in tree growth and a lush growth of mosses that can be superposed on any tropical flora under conditions of excessive cloudiness.

ABOVE THE CLOUDS

Vegetation exists above the clouds in the Solomons, Hawaii and New Guinea and perhaps in other islands, but its organization into plant communities and zones, its prehuman past and its possible future and the location of an effective frost line are all so imperfectly known that few generalizations can be made about them. In Hawaii frosts are not infrequent as low as 4,000 feet and ice formation is appreciable at 7,000 feet.

Much of alpine New Guinea, at least to 12,000 feet, is open and treeless, convincingly interpreted by Lane-Poole 25 as originally forest, but now fire-caused grassland. Here native agriculture is not the chief offender, but fire used as a tool in hunting the wallaby. The higher elevations in Netherlands New Guinea have other zones that are still unknown.

In the Hawaiian Islands the picture is again confused, for here feral grazing mammals have altered the vegetation to an unknown degree and it is doubtful if there are any truly alpine forest species. This paucity of the flora veils the present position of the frost line in the native vegetation. Several thousand feet of elevation are involved, in which Sophora chrysophylla, an important source of fence posts, is the most abundant woody plant. These high regions have probably some of the most intriguing forestry potentialities in all of Oceania.

Forest Influences

While forest influences have long been a matter of abundant fact and fancy, their significance in tropical islands is an unexploited field of research. The fundamental thesis is that the vegetation in the normal course of its development produces certain changes in the microsite and the local environment -- for example, it permits soil accumulation and increasing water infiltration and retention, which in turn, enable the vegetation itself to develop still farther.

Because of water transpired, of absorption and reflection of radiant energy from the foliage canopy and of the resultant effects on the temperature, moisture and saturation deficits of the air above, possible alterations in the regional rainfall are also involved. Animals introduced on small islands may destroy the vegetation, then die themselves, leaving a sandy barren. Rainstorms coming across the ocean may evaporate in the hot air over an island, but yield precipitation on each side. Clouds passing across the mountains may persist over forested coves and forest plantations, but evaporate over anthropically induced savannas and grasslands.

Drizzles may dry up over a roadway, as reported from Martinique in the West Indies, but fall on the forest on either side.

Even a nontechnical observer will be astonished in the tropics at the contrast between a shady, dank, cool, deep- and moistsoiled forest and the scorching rocky man-made desert adjacent.

In short, the magnitude of what vegetation can do to bare land on a tropical island is on a scale for which experience confined to temperate regions does not prepare us. The remedial potentialities of revegetation in low latitudes have been only dimly perceived in research and in practice.

FOREST MANAGEMENT AND USE

VALUE OF THE FORESTS

Timber holds first place among the forest products of New Guinea and Fiji and may do so in the future in the other continental islands, as is generally true on the continents. On the oceanic volcanic islands, however, the native forests bear very few species or individual trees of such size and quality as are suitable for commercial timber and experiments are constantly progressing with introduced plants. The aborigines use wood for house posts, paddles, canoes and sundry minor articles, for which individual trees are selected when needed.

In Hawaii the chief value of the Forest Reserves is in relation to watersheds, because of the demands of the sugar and pineapple plantations and of urban communities and the same probably also holds true for the other high islands under more intensive use.

Where there is mangrove vegetation, it is extensively exploited for fuel and in Fiji is under a 40-year management cycle, as we have seen.

On most of the islands the main use of secondary growth is for firewood and charcoal. For fence posts, highly important in all grazing regions, the naturalized Prosopis chilensis is one of the chief sources of supply. There is no pulp industry of importance, although recent developments in the use of bagasse are significant. Rubber has been tried in the southwestern islands, but not successfully. Minor forest products are of no more than local importance. For the New Guinea native the principal value of the forest lands is as agricultural soil and once the grasses take over, he moves on, or vanishes.

OWNERSHIP AND ADMINISTRATION

Organized forestry, private or governmental, exists in only a few of the island groups. During the Japanese administration of Micronesia, forestry activities were accelerated, with local sawmill operations, minor exports to Japan and a certain amount of reforestation, especially with Casuarina equisetifolia. The present forestry situation in American Samoa and Micronesia has recently been surveyed. New Caledonia, Fiji (where a Forest Department was established in 1938) and Hawaii have, or have had, some government administration, with Hawaii far in the lead.

In Hawaii a Board of Commissioners of Agriculture and Forestry has existed since 1903. Its activities have centered largely around the establishment of reserves almost entirely for watersheds, the elimination of the hordes of feral mammals, fencing, a very elaborate planting programme that has involved a vast number of alien forest species and the controversial practice of introducing game animals. In the early days, forest plantings were sometimes made in total disregard of basic silvical principles and failures were excessive. For a long time, success was gauged on survival and rapid growth and undoubtedly this is a chief factor in the case of a timber species that is to be harvested. For self-perpetuating forests, however, this policy may be very deceptive. An intensive report is much to be desired on all these forest plantings of the past, not only with regard to the survival of planted individuals, but also as to whether different species are capable of spreading naturally. Some species may be found to act as "nurse crops" that is, to prepare the way for the introduction of another semi-natural plant community closer to the original self-perpetuating forest.

UTILIZATION

The original uses of the forest were generally unorganized and unplanned. The natives knew their trees and chose carefully the species and individuals best suited for house posts, canoes, paddles, fishnet floats and other articles to meet their daily needs.

The breadfruit was recognized as one of the most valuable of timber trees (propagated vegetatively and existing in endless clonal strains), but was seldom cut, because its fruits were used for food. Coconut also served numerous purposes. Other trees of the strand were selected and many species in the coastalplain forests of the high islands were discriminately chosen, although reports concerning the same species from different island regions vary markedly. Before World War II organized forestry and forest exploitation can hardly be said to have existed in Oceania. There were a few sawmills on the larger islands, but they served immediate local needs only.

During the war exploitation was stepped up to a high pace to serve the armed forces, especially in the larger Melanesian islands. As in so many tropical regions, the number of species here is great and bewilderingly complex and many of the species have native names only and are unknown to scientific classification. The wartime operations helped to advance our knowledge of the dendrology and physical characteristics of these forests farther than throughout all previous history, but unfortunately most of such information exists only in unpublished reports and in the minds of the foresters who were directly involved.

RESEARCH

The forestry and agricultural organizations are carrying forward only limited programs of basic forestry research. Usually these organizations are under pressure to serve immediate needs and hence their investigations are oriented toward the planting of exotic tree species of proven worth elsewhere and toward the control of actively destructive factors, such as feral mammals.

On the other hand, scientific research in Oceania is beginning to advance rapidly along general lines and this should benefit forestry in the long run.

Forestry in Oceania involves the management of lands originally or potentially covered with trees and such products of forest land as water and wildlife may locally be more important than commercial timber. For these reasons a sound forestry programme must go beyond the range of conventional silvicultural activities. It must include interpretations of the vegetation of pre-European and pre-native times and also serious studies of the remaining samples of natural and semi-natural vegetation, which in some instances have very desirable characteristics. The artificially induced forest types, such as those from plantings, should also be investigated with regard both to their continuing or self-perpetuating characteristics and to their values in other respects than for timber production. Because of the extraordinary complexity of the native and introduced floras and because of the wide range of distinctive environmental conditions within relatively small areas, forestry in Oceania offers unsurpassed challenges to the ingenuity of the practical researcher.

4

Forest Plantations and Biodiversity a Fresh Perspective

Forest plantations covered 187 million hectares in 2000, of which Asia accounted for 62 per cent. The forest plantation area represents a significant increase from the 1995 estimate of 124 million hectares. The reported new annual planting rate is 4.5 million hectares globally, with Asia and South America accounting for 89 per cent. About 3 million hectares are estimated to be successful. Globally, half the forest plantation estate is for industrial end-use, one-quarter for non-industrial end-use and one-quarter not specified. Globally, the main fast-growing, short-rotation species are in the genera *Eucalyptus* and *Acacia*. Pines and other coniferous species are the main medium-rotation utility species, primarily in the temperate and boreal zones.

The potential for forest plantations to partially meet demand from natural forests for wood and fibre for industrial uses is increasing. Although accounting for only 5 per cent of global forest cover, forest plantations were estimated in the year 2000 to supply about 35 per cent of global roundwood. This figure is anticipated to increase to 44 per cent by 2020. In some countries forest plantation production already contributes the majority of industrial wood supply. There is increasing interest in development of forest plantations as carbon sinks; however, failure to resolve international debates on legal instruments, mechanisms and monitoring remains a serious constraint.

In developing countries about one-third of the total plantation estate was primarily grown for woodfuel in 1995 - although it should be noted that planted trees on farmland, in villages and homesteads and along roads and waterways contribute significantly to fuelwood supplies, enabling the demand to be met in most instances.

New forest plantation areas were reported as being established globally at the rate of 4.5 million hectares per year, with Asia and South America accounting for more new plantations than the other regions. Of plantations established, about 3 million hectares per year were estimated as being successful. Of the estimated 187 million hectares of plantations worldwide in 2000, Asia had by far the largest area. In terms of composition, *Pinus* spp. (20 per cent) and *Eucalyptus* spp. (10 per cent) remain dominant worldwide, although the overall diversity of species planted was shown to be increasing. Industrial plantations account for 48 per cent, non-industrial plantations for 26 per cent and plantations for unspecified use for 26 per cent of the global forest plantation estate.

The results of the plantation assessment were the first global estimates with a uniform definition of forest plantations and can therefore not be directly compared to previous estimates. FRA 2000 country statistics on plantations may also differ from those reported in prior FAO publications, partly because of changes in definitions. Countries participated directly in the assessment, providing technical documentation and supporting analysis and validating the results generated by FAO. Several experts around the world were enlisted to provide detailed information on various aspects of the plantation situation in the form of special studies.

Between the extremes of afforestation and unaided natural regeneration of natural forests, there is a range of forest conditions in which human interventions occur. European forests have long traditions of human intervention in site preparation, tree establishment, silviculture and protection; yet these are not always defined as forest plantations. The traditional forest plantation concept tends to be applied to single species, uniform planting densities and even age classes. Terms such as "natural forest under management" or "assisted natural regeneration" are applied to stands of indigenous species in more heterogeneous management mechanisms in Europe and other industrialized temperate and boreal countries.

METHODS

The area of existing forest plantations would ideally all have been derived from statistically designed inventories of forest plantations or statistics for planted areas reported by planting agencies or appearing in national reports. However, information also comes from many other sources including nursery production, seedling distribution and estimates derived from the goals of planting programmes. The vast range of agencies, industries and non-governmental organizations within countries engaged in planting programmes made the comprehensive collection of all relevant source documents a major logistical exercise. For FRA 2000, over 800 source documents were analysed to derive the forest plantation estimates. In most developing countries a national clearinghouse for collecting information on plantations is either lacking or ineffective owing to the enormity of the task and limited resources.

Data Collection

To retrieve the source documents for the plantation study, FAO made formal requests to all developing countries, some of which contributed the necessary materials. Most of the reports were collected directly by FAO staff during FRA 2000 workshops and visits to national ministries. For consistency FRA 2000 prepared guidelines and questionnaires for the collection of forest plantation statistics in which the objectives, scope, definitions, sources of data and templates for specific data collection were supplied to each country. Parameters requested included:

- Total estimated forest plantation area, 2000;
- Annual area of new plantations;
- Species groups: broadleaf, conifer, non-forest like, African oil palm (*Elaeis guineensis*), coconut palm (*Cocos nucifera*) bamboo or unspecified;
- Purpose and end-use objective of forest plantations: industrial (producing wood or fibre for supply to wood processing industry) or non-industrial (fuelwood, soil and water protection);
- Ownership: public, private, other (e.g. traditional, customary) or unspecified.

Other data requested in the guidelines, which proved difficult for countries to provide by species group, included age class distribution; end-use by forest product (industrial plantations); growth and yield (mean annual increment); standing volumes; and rotation lengths. Despite the absence of these data, FRA 2000 is the most comprehensive forest plantation resources assessment that has been carried out.

In previous assessments of forest plantation resources, plantation data were available up to the reference year for most countries, since the reporting followed the reference year. In FRA 2000, the reference year was 2000, so if data were not available to that date, then existing area and annual planting data were used to extrapolate the necessary information. For the few countries that have no data sets since 1990, the rate of planting in preceding years and future planting programmes were considered in projections to the year 2000.

FAO also enlisted the assistance of several experts around the world to make specific technical contributions on the forest plantation situation in the 1990s. These studies constituted an important part of the global results as well and complemented the country information.

Analysis and Interpretation

The quantity and quality of forest plantation data provided is dependent upon the capacity of the national forest inventory systems to collect and analyse data and to adjust the information to conform with global and regional reporting parameters. In many developing countries there is a lack of institutional capacity to carry out periodic national forest inventories, so data can be incomplete, inconsistent, outdated and of variable reliability. Because of this, it was necessary to derive and in some instances to verify forest plantation statistics through desk research using available country reports. All sources of country data were referenced and made available in a transparent manner.

In addition, regional and national focal persons were appointed to assist in the forest plantation data collection, to ensure that the latest data were available and to maintain coordination and communication between FRA 2000, FAO regional offices and each participating country. On completion of the data sets, a formal verification process was undertaken with each participating country.

Purpose and ownership within the global forest plantation estate Purpose and ownership of forest plantations vary markedly among regions. Industrial plantations provide the raw material for wood processing for commercial purposes, including timber for construction, panel products and furniture and pulpwood for paper. In contrast, non-industrial plantations are aimed for example at supplying fuelwood, providing soil and water conservation, wind protection, biological diversity conservation and other non-commercial purposes.

In many countries, particularly in the developing world, the end purpose of the plantations is not clearly defined at the outset. In some of these cases, valuable tree resources are established which coincidentally match future needs. However, in other cases the lack of planning may result in plantations that have little commercial value and a low potential for local use.

SELECTED GLOBAL TRENDS, 1980-2000

Comparisons

FRA 2000 country statistics on plantations may differ from those reported in prior FAO publications partly because of changes in definitions. For example, rubber plantations were not previously considered as forest plantations but are included in FRA 2000 plantation data. Previous assessments also used regional reduction factors to indicate the successful proportion of plantations remaining after establishment. The FRA 2000 assessment applied reduction factors according to the best available data from each country independently. There have also been changes in the information base from which the estimates were derived. The statistics now include data from many industrialized countries, none of which were included in the prior global assessment reports. Despite these differences, comparison of FRA results from each decade allows analysis of some trends including planting rates, genera, areas and purpose.

IMPACTS OF THE FOREST PLANTATION ESTATE

The potential for forest plantations to partially meet demand for wood and fibre for industrial uses is increasing. According to FRA 2000, the global forest plantation area accounts for only 5 per cent of global forest cover and the industrial forest plantation estate for less than 3 per cent. However, as an indication only, forest plantations were estimated in the year 2000 to supply about 35 per cent of global roundwood and an increase to 44 per cent anticipated by 2020. If plantation development is targeted at the most appropriate ecological zones and if sustainable forest management principles are applied, forest plantations can provide a critical substitute for natural forest raw material supply. In several countries industrial wood production from forest plantations has significantly substituted for wood supply from natural forest resources. Forest plantations in New Zealand met 99 per cent of the country's needs for industrial roundwood in 1997; the corresponding figure in Chile was 84 per cent, Brazil 62 per cent and Zambia and Zimbabwe 50 per cent each. This substitution by forest plantations may help reduce logging pressure on natural forests in areas in which unsustainable harvesting of wood is a major cause of forest degradation and where logging roads facilitate access that may lead to deforestation.

Forest plantations also provide additional non-wood forest products, from the trees planted or from other elements of the ecosystem that they help to create. They contribute environmental, social and economic benefits. Forest plantations are used in combating desertification, absorbing carbon to offset carbon emissions, protecting soil and water, rehabilitating lands exhausted from other land uses, providing rural employment and, if planned effectively, diversifying the rural landscape and maintaining biodiversity.

Not all forest plantation development has positive economic, environmental, social or cultural impacts. Without adequate planning and without appropriate management, forest plantations may be grown in the wrong sites, with the wrong species/provenances, by the wrong growers, for the wrong reasons.

Examples exist where natural forests have been cleared to establish forest plantation development or where customary owners of traditional lands may have been alienated from their sources of food, medicine and livelihoods. In some instances poor site/species matching and inadequate silviculture have resulted in poor growth, hygiene, volume yields and economic returns. In other instances, changes in soil and water status have caused problems for local communities. Land use conflicts can occur between forest plantation development and other sectors, particularly the agricultural sector.

The negative impacts of forest plantations can draw the focus away from the fact that forest plantation resources are totally renewable and can be economically, socially, culturally and environmentally sustainable with prudent planning, management, utilization and marketing.

SELECTED FOREST PLANTATION TOPICS

Mean Annual Volume Increment (MAI) of Select Industrial Species

On average *Eucalyptus* and *Pinus* species, which dominate industrial plantations in developing countries, have similar MAIs of 10 to 20m^3 per hectare per year. However, many of the popular species of both genera frequently achieve much faster growth rates. Thus *Eucalyptus grandis,* which is the most widely planted *Eucalyptus* species, can achieve 40 to 50 m^3 per hectare per year and in very exceptional conditions with advanced tree improvement 100m^3 per hectare per year. Other widely planted tropical hardwoods including *Casuarina equisetifolia, Casuarina junghuhniana, Tectona grandis* and *Dalbergia sissoo* have MAIs of less than 15m^3 per hectare per year and frequently under 10 m^3 per hectare per year.

Climate and site have a very large impact on growth rates. The humid tropics and more fertile sites are more conducive to higher growth rates than locations with long dry seasons or infertile or degraded soil. Teak on many sites in India, for example, frequently has an MAI of 4 to 8 m^3 per hectare per year, partly because of drought combined with poor soils. Some species such as *Gmelina arborea* and some of the *Eucalyptus* species are very site sensitive. *Pinus* spp., in contrast, generally tolerate adverse conditions better and are more flexible with respect to site.

Both tree breeding and silviculture have improved growth rates. Good examples are *Eucalyptus grandis* and *E. urophylla* in Brazil and *Pinus radiata* in some countries of the Southern Hemisphere.

Advanced silviculture typically includes improved nursery and establishment techniques such as good site preparation, weed control and judicious use of fertilizer. It has been suggested that growth of teak, for example, could be doubled in Kerala, India and Bangladesh and increased sixfold in Indonesia by adopting these practices.

With coppice species productivity varies with rotation, the first and second coppice rotations usually being more productive than the seedling one. The growth patterns vary among species. For example, very fast growing species such as *Gmelina arborea* can reach a peak MAI in less than 10 years, while *Pinus caribaea* var. *hondurensis* grown in Trinidad reaches maximum MAI at about 25 years and *P. radiata* at over 40 years. With *Cupressus lusitanica* in Costa Rica, the MAI maximum is reached at about 30 years.

Rotation lengths can reflect both end-use and economics. Many fast-growing *Eucalyptus, Acacia* and *Casuarina* species and *Gmelina arborea* are grown on short rotations of under 15 years as they are used primarily for pulp or woodfuel. Usual rotations in Kenya for *E. grandis* are 6 years for domestic woodfuel, 7 to 8 years for telephone poles and 10 to 12 years for industrial woodfuel. In Brazil this species is largely grown for pulp or charcoal on 5 to 10 year rotations. Species being grown for high-value sawlogs usually have longer rotations; teak (*Tectona grandis*) is grown on 50 to 70 year rotations and high-value conifers such as *Araucaria angustifolia* on 40 year rotations. Generally pines are grown on medium-length rotations of 20 to 30 years, unless grown solely for pulpwood, when shorter rotations may be adopted.

Modelling of growth, rotations, harvest yields and product mix by species is important for decision-making in forest management. One of the major obstacles to model development for planners and managers is a lack of suitable data. Data can come from a range of sources, including temporary and permanent sample plots and experiments. Experiments and protocols for obtaining data need to be carefully designed so that reliable information is obtained over the complete range of conditions to which the model is to apply. Tree Growth and Permanent Plot Information System (TROPIS), sponsored by the Centre for International Forestry Research (CIFOR), seeks to coordinate and improve access to tree growth information.

A growth and yield model developed for *Pinus elliottii* plantations in coastal Zululand, South Africa, predicts height, basal area, total stem and merchantable volume and stocking, with age on a stand level for harvest planning. In New Zealand several simulation models have been developed for *P. radiata* which predict similar variables but also include wood quality, harvesting and marketing aspects, which make it possible to link the silvicultural options to industrial use.

Sustaining Productivity

It is possible not only to sustain but also to increase productivity in successive rotations. This requires clear definition of the end-use objective for forest plantation development and a holistic view in their management. There is a need to integrate strategies for genetic improvement programmes, nursery practices, site and species/provenance matching, appropriate silviculture (site preparation, establishment, weeding, fertilizing, pruning, thinning), forest protection and harvesting practices with prudent management. New Zealand and the southern United States have shown that substantial gains can be made by adopting this holistic approach.

In developing countries where resources may be constrained, highly technical solutions may not be essential but it is critical to get the fundamentals correct: careful species and provenance choice, good nursery stock, site preparation, planting techniques, weed control and, less frequently, fertilizer inputs. Once fast-growing, uniform plantations have been established, later silvicultural tending may become increasingly important, depending on the end-use objective.

Current evidence suggests that plantation production can be sustainable if foresters implement prudent genetic and silvicultural tree improvement programmes and sound management practices. There has been, however, limited long-term research on the subject; there are few definitive studies, limited to few species.

In one of the most promising studies, with *Pinus patula* in Swaziland grown intensively on about 15 year rotations, site productivity was maintained or increased over three rotations. The question of declining growth in teak (*Tectona grandis*) plantations in Indonesia and India remains unclear.

How forest plantations are managed affects the chemical and physical properties of the soils and site. However, only recently have long-term studies been undertaken to evaluate these critical factors or processes.

The methods adopted for site preparation (ripping, ploughing, scarifying, bedding, windrowing, controlled burning), establishment (manual, mechanical), weeding (manual, chemical, mechanical), fertilizer application, pruning and thinning (manual and mechanical, for commercial to waste), forest protection and harvesting (manual, mechanical, clear-fell or selection) all affect the pool of nutrients in the ecosystem.

Interference with the drainage, litter and recycling of organic matter and change in the physical conditions of soils during these operations are critical to long-term sustainability. Because of litter recycling and the rapid development of tree roots, plantations are used for rehabilitation of fragile and degraded lands prone to soil erosion and excessive water runoff.

Tree plantations often have higher evapotranspiration rates than grassland or agricultural crops and thus change the hydrology of the site. This can be either beneficial (for example by reducing salinity problems in some dryland conditions) or detrimental (if it reduces water required for other uses).

The rare studies of changes of productivity between rotations have concluded that negative changes have primarily been due to inappropriate or inadequate management practices or weed invasions rather than a result of the plantations themselves.

Burning and excessive cultivation in site preparation, soil compaction from mechanical operations, inappropriate harvesting techniques and poor forest protection can contribute to loss of nutrients and soil erosion, with a resultant loss in productivity of forest plantation sites. This cannot be addressed solely by addition of fertilizer, but by the adoption of the whole range of tree improvement, silviculture, protection and harvesting techniques in an integrated forest management strategy.

Valuable Hardwood Plantations

Long-rotation, slow-growing but valuable hardwood species have special technical properties, such as strength, natural durability, hardness and easy machining and appearance (grain, figure, texture, colour and other aesthetic qualities) that make them suitable for high-value end-uses such as furniture.

These high-grade hardwoods contrast with short-rotation, fast-growing, lesser-quality woods used for woodfuel, pulpwood or reconstituted products and less demanding building timbers. In tropical countries teak (*Tectona grandis*), mahogany (*Swietenia* spp.) and rosewood (*Dalbergia* spp.) are the main hardwood plantation species, while in temperate countries oak (*Quercus* spp.), ash (*Fraxinus* spp.), cherry (*Prunus* spp.), walnut (*Juglans* spp.), tulipwood (*Jacaranda* spp.) and hard maple (*Acer* spp.) predominate.

Table. Characteristics of Valuable Hardwoods used in Tropical Areas

Use categories	*Desirable wood properties*	*Main end-uses*	*Matching valuable hardwood species*	*Comments*
Decorative timbers	Appearance, consistent quality, dimensional stability, durability, good machining, staining and finishing properties	Quality furniture and interior joinery	*Tieghemella* spp.; *Entandophragma cylindricum, Chorophora* spp., *Aucoumea klaineana, Afrormosia* spp., *Entandophragma utile, Mansonia* spp., *Lovoa* spp., *Khaya* spp., *Swietenia* spp., *Dalbergia* spp., *Aningeria* spp.	Highest value, competitio n from temperate hardwoods and MDF
High to very high-density timbers	Appearance, strength, high natural durability, availability in large sizes	Principall y in constructi on	*Dipterocarpus* spp., *Lophira* spp., *Chlorophora* spp., *Ocotea rodiaei*	Small share of total tropical timber use
Low to medium-density utility timbers	Appearance, clear grain, natural durability, good machining properties	External joinery, shop fittings, medium-priced furniture	*Shorea* spp, *Hevea brasiliensis, Terminalia* spp., *Heritiera* spp.	Most commonly used, prone to competitio n from substitutes

Because many valuable hardwood species are difficult to establish because of their ecological requirements or disease or insect susceptibility, focus has been on the easier species to grow, including teak (*Tectona grandis*), Indian rosewood (*Dalbergia sissoo*) and mahogany.

In 1995 the global areas of these species were 2 254 000, 626 000 and 151 000 ha, respectively.

They accounted for about 10 per cent of total hardwood plantations in the tropics. More than 90 per cent of teak plantations were located in Asia, mainly in Indonesia, India, Thailand, Bangladesh, Myanmar and Sri Lanka. About 95 per cent of rosewood plantations are located in India and Pakistan.

The largest mahogany plantations are located in Indonesia and Fiji, which together make up about 80 per cent of the established area.

The market preference for large piece sizes, slow growth and very long rotation lengths combine to reduce the attractiveness for commercial investment in these species.

This is only partially counteracted by their value. The low return on capital investment, coupled with the long wait period for this return, has made it difficult to interest private investors without supportive, secure and stable government policies.

As markets demand a continuity of supply, plantations need to be on a sustainable scale within a region. Some of the less common species are not known in the marketplace. Other potential market problems are that the timber may be wrongly associated with tropical deforestation and changing fashions that often occur with decorative timbers. Niche marketing is important for valuable hardwoods.

Projections for supplies of timber from existing valuable hardwood plantations indicate that because of the age class distribution and long rotations there will not be a significant increase in supply in the next 20 years. Future promotion of quality hardwood plantations needs to emphasize choice of species with versatile end-uses, market research and development to hold on to niche markets and maintained high standards from production to marketing.

Careful site selection, use of high-quality planting materials of superior genetic origin and good silviculture are important. Planting programmes should be economically viable, environmentally appropriate and socially desirable. Incentives may also be necessary to stimulate private investment because of the long rotations.

Even though valuable hardwood plantations have the potential to reduce the pressure on natural forests, they will not prevent deforestation resulting from agricultural encroachment.

The supply of large quantities of high-value timber could perhaps undermine the value of natural forest stands and so lead to more rapid destruction. Hence it is advisable, where possible, to manage plantations and forest resources and forest products in a complementary manner.

Plantations and Wood Energy

Woodfuels from plantations or natural or semi-natural forests are particularly important in developing countries, providing about 15 per cent of their total energy demand. Woodfuel provides about 7 per cent of energy demand for the world as a whole and in industrialized countries only 2 per cent. Woodfuel provides more than 70 per cent of energy needs in 34 developing countries and more than 90 per cent in 13 countries. Woodfuel makes up about 80 per cent of total wood use in developing countries and about 89 per cent in Africa

The prediction of a woodfuel crisis in developing countries in the 1980s was based largely on looking at supply and demand from forest plantations and natural forests. The reaction to the expected woodfuel crisis was to plant trees for this purpose, often in the form of traditional plantations.

Many programme failures resulted from lack of appreciation for the complexities of bioenergy supply and demand, failure to take into account social aspects and people's needs and poor programme structures. The importance of planted trees on farmland, in villages and homesteads and along roads and waterways as a source of woodfuel supply was underestimated.

Rural communities harvest stems, branches, stumps, twigs, leaves and litter for woodfuels in chronic woodfuel supply areas. In these instances the nutrient recycling process is broken, resulting in degradation of forest plantation sites. In many rural communities in developing countries, woodfuel is considered a public free good, to be foraged from public natural and plantation forests. Often women and children collect the woodfuel at little or no cost. As a result, the growing of private forest plantations specifically for woodfuel, in which development costs and rotation cycles are involved, can be a foreign concept.

Asian studies show that forest-based supply can range from 13 per cent in the Philippines to as high as 73 per cent in Nepal. In many countries less than 50 per cent of fuelwood is from forests.

Globally, non-industrial forest plantations in 1995 were estimated to cover about 20 million hectares. This was almost 17 per cent of the world's total plantation area in 1995. A significant proportion of these plantations were planted for woodfuel and 98 per cent were in developing countries. These plantation figures do not account for trees planted outside the forest on farms or in villages, etc., nor do they consider plantations that were considered agricultural plantations, such as *Hevea* or palm plantations.

In developing countries about one-third of the total plantation estate was grown primarily for woodfuel in 1995. Three-quarters of these plantations were in Asia, where they accounted for 60 per cent of total plantation production. In Latin America more than half of plantation production went to woodfuel; in Africa and Oceania a larger proportion of plantation production was as industrial wood. However, plantations, in general, provided only a small proportion of total woodfuel used. Uruguay is an interesting exception.

Production of woodfuel from plantations currently makes only a small contribution to energy requirements, although it is very important in some localities and countries. Plantations currently supply 5 per cent of woodfuel. Production from these non-industrial plantations is likely to double over the next 20 years, even with little expansion in area, because the age class distribution is heavily concentrated in young plantations. In an optimistic scenario where planting continues at the same rate as in the past ten years and then gradually declines, a 350 per cent increase in woodfuel production would be anticipated by 2020. By-products from wood-using industries will also contribute to increased fuelwood supply. The situation is less positive in Africa, where for a few countries declines are projected in plantation-based woodfuel production.

New Sources of Fibre

Since FRA 1990, advances in wood utilization technology have resulted in increasing importance of new sources of fibre - rubber (*Hevea brasiliensis*), coconut palm (*Cocos nucifera*) and African oil palm (*Elaeis guineensis*) - especially in the Southeast Asian subregion.

These species account for 9.7, 12.0 and 6.0 million hectares of plantations, respectively. All grow in the humid tropics. In terms of plantation area, Asia has 92 per cent of the world's rubber, 86 per cent of the world's coconut palm and 78 per cent of the world's African oil palm. Indonesia,

Thailand and Malaysia have almost three-quarters of the rubber plantations; Indonesia and the Philippines have about half the coconut resources; and Malaysia has 55 per cent of the oil palm resource. All three species are grown principally for other products rather than wood, so when overmature they are available for fibre-based industries at minimal cost.

Rubberwood is harvested when latex productivity declines and yields 100 m^3 per hectare of roundwood, but recovery for lumber is only 25 to 45 per cent because of poor form and small size. Most of the planted stands in Southeast Asia are owned by smallholders and are geographically dispersed, with poor accessibility and poor-quality stems. Currently the major proportion of industrially utilized rubberwood comes from large-scale plantations. Quality furniture, parquet, panelling, reconstituted panels, general utility timber and woodfuel, including charcoal, are made from rubberwood.

However, the rubberwood must be processed within days of harvesting to minimize sapstain attack. The most developed downstream industries are in Malaysia, where the production of sawn rubberwood timber rose from 88 000 m^3 in 1990 to 137 000 m^3 in 1997 and medium density fibreboard (MDF) production from rubberwood reached 1.16 million cubic metres per annum by 1999. Exports of rubberwood furniture have grown from about US$74 million in 1991 to US$683 million in 1998. Rubberwood has become a substitute for light tropical forest hardwoods. Its acceptance as a sustainable plantation-grown, environmentally friendly timber has given it wide appeal.

Coconut palms are harvested as the copra yields decline (beyond 60 years) and yield 90 m^3 per hectare of coconut wood. Coconut palm has variable properties and is intrinsically difficult for conversion but can yield a relatively low-cost, general-utility timber for construction, panelling, stairs, door jambs, furniture, flooring and power poles.

In 1993 Indonesia had 65 million cubic metres of overmature coconut stems which needed disposal before replanting. There is increasing interest in this raw material in European and North American markets. It is unlikely to replace conventional timber, but likely to find its way into niche markets. It will continue to be used as a low-cost construction timber.

Oil palm plantations are harvested for fibre beyond the 25 to 30 year rotations and yield about 235 m^3 per hectare. It is estimated that over 1.6 billion cubic metres of fibre will be available in the years to come from established resources in Southeast Asia. From 1996 to 1999 the area increased by 18 per cent. In Malaysia the area has increased by 3 million hectares in the past 30 years.

Most oil palm plantations (unlike rubber and coconut) in the main growing countries, Malaysia and Indonesia, are managed by plantation companies or cooperatives. Oil palm by-products such as kernel shells, pressed fibres and empty fruit bunches are currently used in heat generation at the extraction plants. Water in the stems can reach five times the weight of dry matter.

The high moisture content as well as the high amounts of parenchyma tissue rich in sugar and starches make conversion into quality forest products a challenge. An MDF plant in Malaysia is currently being planned to utilize oil palm stems.

Plantation Substitutes for Natural Forest Products

With growing concerns about the status and loss of natural forests, the rapid expansion of protected areas and large areas of forest unavailable for wood supply, plantations are increasingly expected to provide substitutes for products from natural forests, particularly in Asia and the Pacific.

In Asia and the Pacific it is estimated that 52 per cent of natural forests are not available for wood harvest because they are inaccessible or uneconomic to exploit. Of the unavailable forest in the region, it is estimated that about 38 per cent is legally reserved. In addition, logging bans have been imposed on large areas of natural forest covering about 10 million hectares. The reasons for these bans vary but were related to deforestation and forest degradation causing environmental problems in Thailand, the Philippines and China and to conservation requirements in Sri Lanka and New Zealand.

As a result of the net effect of deforestation and removal of natural forests from wood production, some areas in the Asia and the Pacific region have wood deficits and roundwood harvesting is exceeding sustainable levels of cut. The worst affected areas are South Asia and insular Southeast Asia, with continental Southeast Asia also under strain. In contrast, New Zealand has surplus plantation wood available for export.

Of six examples studied in the Asia and the Pacific region, New Zealand is more than self-sufficient in wood production based on plantations. In China and Viet Nam, the importance of plantations will increase as planted resources mature. There have been serious problems with implementing plantation development programmes in Sri Lanka, the Philippines and Thailand. In Sri

Lanka, India and elsewhere in the tropics, trees outside the forest are playing a critical role in roundwood and woodfuel supply.

Most countries in the region are becoming importers of wood, with imports expected to rise. Sometimes logging bans have shifted the problem to other countries. Problems with acquiring large areas of land in some countries make it difficult to implement industrial plantations. In the Philippines, Thailand and Viet Nam there have been social conflicts with local indigenous people or between traditional forest use and development, as well as between the rich and the poor. Sometimes incentives and the development of social forestry programmes are being used to help resolve such problems.

While it is clear that plantations will have an increasingly significant role in substituting products from natural forests, the impact will be felt on a case-by-case basis as governments and investors determine where and how plantations can be technically, economically and socially feasible as well as environmentally friendly.

In the near term, plantations in Asia and the Pacific can make a contribution but cannot replace harvests from natural forests. It is likely that both in the region and globally the current pace of industrial plantation development will barely keep pace with losses from deforestation and transfer of natural forests to protected status. While it would be theoretically possible, actual plantation development is at present not sufficient to offset both growing consumption and declining harvest from natural forests.

Plantations and Carbon Sequestration

In the past ten years, the development of forest plantations as carbon offsets has evolved towards a market mechanism, although an organized market with carbon prices defined according to supply and demand forces is still a long way off. The adoption of the Kyoto Protocol in 1997 triggered a strong increase in investment in plantations as carbon sinks, although the legal and policy instruments and guidelines for management are still debated.

A number of countries have already prepared themselves for the additional funding for the establishment of human-made forests. The 1997 Costa Rica national programme was the first to establish tradeable securities of carbon sinks that could be used to offset emissions and the first to utilize independent certification insurance.

To date, greenhouse gas mitigation funding covers about 4 million hectares of forest plantations worldwide. The recognition of afforestation and reforestation as the only eligible land use, land use change and forestry activities under the Clean Development Mechanism of the Kyoto Protocol, as agreed in Bonn during the second part of the Sixth Conference of the Parties to UNFCCC in July 2001, will lead to a steep increase in forest plantation establishment in developing countries. The sink decision of the Bonn Agreement is expected to funnel additional funds into forest activities in developing countries and thus to strengthen the international efforts in this field.

However, it will also require a monitoring and verification system to ensure that these plantations will not be established at the expense of the local population or efforts to conserve biological diversity. Thus the decisions taken in Bonn to make the Kyoto Protocol ratifiable will also bear new challenges for forest plantation development.

Trees Outside the Forest

The significance of trees outside the forest (TOF) can be observed in several contexts. In countries with low forest cover, TOF resources constitute the main source of wood and non-wood "forest" products, even though trees may be so scattered that the maps produced by FRA 2000 indicate that no forests exist.

Trees are found on agricultural lands, in densely populated areas, in fruit-tree plantations and in home gardens, which often cover a large proportion of the land. In urban areas trees provide important aesthetic and environmental services in addition to providing shade and greatly increasing the livability of cities. Communities, farmers and herders who do not have access to forests diversify their production and protect their land by maintaining various tree systems on their farms.

Deforestation has been mapped and quantified, but very little is known about the fate of land formerly under forest; forest clearing is often followed by the establishment of production systems of which trees are an integral part. Not much is known about the dynamics of trees on farmlands and their corresponding contribution to the production of wood and other products and services. Similarly, little is known about changes in tree cover in fields and urban systems. Knowledge of trees outside the forest comes mostly from local studies on agroforestry, sylvipastoralism and urban, social, community or rural forestry.

This widespread and multipurpose resource, familiar to farmers but poorly defined by managers and mostly absent from official statistics and development policies, needs to be better assessed and known.

Growing populations, shrinking forests and degraded ecosystems all suggest that trees outside the forest are destined to play a larger local and global role in meeting the challenges of resource sustainability, poverty reduction and food security. Trees outside the forest relieve the pressure on forest resources, conserve farmland, boost agricultural productivity, blunt the harmful impact of urban growth on the environment, increase food supplies, provide income and in general make valuable contributions to food security.

FRA 2000 did not undertake a global assessment of trees outside the forest, mainly because of resource limitations; nor has there ever been a comprehensive global assessment of trees outside the forest and their products. However, a number of studies have been carried out for specific sectors or geographic areas, often with an emphasis on their economic contributions. This chapter provides a summary of selected studies and discusses the practical and conceptual difficulties related to a comprehensive global assessment.

The chapter responds to the concern expressed by the Expert Consultation on Forest Resources Assessment 2000 regarding the lack of information on TOF.

DEFINITION OF TREES OUTSIDE THE FOREST

Trees outside the forest are defined by default, as all trees excluded from the definition of forest and other wooded lands. Trees outside the forest are located on "other lands", mostly on farmlands and built-up areas, both in rural and urban areas. A large number of TOF consist of planted or domesticated trees. TOF include trees in agroforestry systems, orchards and small woodlots. They may grow in meadows, pastoral areas and on farms, or along rivers, canals and roadsides, or in towns, gardens and parks. Some of the land use systems include alley cropping and shifting cultivation, permanent tree cover crops (e.g. coffee, cocoa), windbreaks, hedgerows, home gardens and fruit-tree plantations.

Classification of trees outside the forest presents certain difficulties. There are existing classifications for agroforestry, but none applicable to all trees outside the forest. For practical reasons, the FRA 2000 definition of "forest" combines aspects of both land cover and land use. This approach creates difficulties not only for classification of forest, but also for classification of TOF.

In a study on data-gathering on TOF in Latin America, where classification was primarily based on land use criteria, separating land use and land cover aspects was found to be a main source of misinterpretation. There was a possibility of confounding coffee plantations and trees in pasture with forest, given their high density.

This clearly shows some of the problems involved in establishing a simple and reliable *a posteriori* classification.In France, the National Forest Inventory (IFN) and Teruti Land Use Study-have begun to attempt coordinating classifications of trees outside the forest. The objective is eventually to use the annual Teruti data to update the IFN ten-year data, with a single national nomenclature as a possible end result.

FUNCTIONS AND CHALLENGES

In industrialized countries, farmers list shade and shelter, soil protection and improvement of the landscape and rural environment as their main reasons for growing trees. In the tropics, farmers grow woody species for food security and subsistence. Trees outside the forest are a major source of food. Livestock fodder produced by TOF can be a matter of life and death in semi-arid or mountainous areas. Fuelwood remains the prime source of energy in developing countries, representing up to 81 per cent of the wood harvest (FAO 1999). In contrast, in the industrialized countries, fuelwood accounts for less than 10 per cent of total fuel consumption (FAO 1998). Very few studies have reported on overall fuelwood output from stands and single trees outside the forest, but agroforestry systems and orchards are known to provide a large part of the resource.

Trees outside the forest have an important ecological role. Planted trees and shrubs in fields help to check runoff and erosion and control flooding, as well as helping to purify water and protect against wind. Trees lining rivers and streams help to maintain biodiversity, providing spawning beds for fish and shellfish and shade which reduces eutrophication.

The unique role of trees in soil protection and conservation, checking wind and water erosion and maintaining soil fertility is universally acknowledged. Also important are the cumulative benefits of trees on smallholdings to soil and water conservation, in particular in the larger context of mountain watershed management; their positive impact on climate; and their role in buffering the effects of desertification and drought.

RESULTS OF SELECTED STUDIES

In spite of the limits of data at the regional or global level, a number of local initiatives have been carried out. The approaches of the various studies differ in accordance with the purpose and scale of the analysis. Few studies use methods resembling the conventional forest inventory.

Many studies rely on existing literature or estimates drawn from surveys and interviews. The quantification of products is often based on different parameters, such as estimates of global output, marketed output, observed or potential productivity or economic value. Thus the reliability of the results is uncertain.

The following are the results of some national initiatives that have assessed trees outside the forest. In Kerala, the most densely inhabited state of India, a study estimated that of the total annual production of 14.6 million cubic metres of wood in the state, about 83 per cent was from homesteads (house compounds and farmlands), 10 per cent from estates (plantations of rubber, cardamom, coffee and tea) and only about 7 per cent from forest areas 26.6 per cent of the state area is under forest cover.

Trees outside the forest met about 90 per cent of the fuelwood requirements of the state. Fuel from coconut trees alone, including both wood and non-wood materials (pruned and fallen), constituted about 70 per cent of the total fuelwood supply.

A study in Haryana State in India, an intensively cultivated state with about 3.8 per cent of its area classified as forest land but only about 2 per cent under actual forest cover showed that farm forestry (trees along farm bonds and in small patches up to 0.1 ha) accounted for 41.2 per cent of the total growing stock of wood. Multiple tree rows along roads and canals accounted for 13 per cent and 9.6 per cent, respectively; village woodlots for 24 per cent; and block plantations of less than 0.1 ha for 10.6 per cent.

In Morocco, where forest cover is less than 5 per cent of the land cover and other wooded lands only 7 per cent, nearly 20 per cent of the land may be occupied by trees outside the forest, namely as wooded pasture (84 per cent) and fruit-tree plantations (12 per cent) (Rosaceae, citrus, olives trees, palm trees, walnut trees, fig trees, almond trees).

Fruit production has an important place in the national economy. It is noteworthy that even when a forest is largely destroyed, the carob is one of the few species traditionally conserved, as it is highly appreciated by farmers for multiple purposes, providing both fodder and income from the sale of its fruit for export.

However, there are no reliable data on the distribution and potential of this "forest" resource, which is of interest for farmers, herders, concessionaires and the government and distributed on agricultural and forest lands. In the Sudan, the National Forest Inventory has undertaken a national land use inventory to provide area and volume statistics for planning at the subnational and national levels. The inventory was designed to provide preliminary estimates regarding products other than the traditional fuelwood and timber, such as the amount of gum, fruit or nuts that can be collected and the distribution of non-wood species of interest.

In Costa Rica, the Tropical Agricultural Research and Higher Education Centre (CATIE), in collaboration with Freiburg University, Germany, is developing a regional methodology for Central America to assess tree resources outside the forest.

A mix of satellite remote sensing, aerial photos and ground sampling is used to address the complexity of the resource and to allow dynamic monitoring of resources at the national and regional levels data-gathering on TOF in eight Latin American countries. None of these countries had established a database and the search for information was multisectoral. The statistics on land cover and land use gave some idea of the relative importance of trees outside the forest.

In Kenya, extensive tree planting on farmlands was promoted in the 1970s and 1980s, with land tenure security as a major incentive. There is an increasing trend of tree cover and species diversification on privately owned farms. Assuming that the present rate of increase in tree planting will continue, it was estimated that farms produced about 9.4 million cubic metres of wood in 2000 and will produce about 17.8 million cubic metres in 2020. Their share of the total wood produced in the medium- and high-potential districts was projected to increase to 80 per cent in 2020. Indeed, while natural stands of trees have declined there has been a corresponding increase in tree planting in much of the densely populated plateaus of Kenya. As natural forests are reduced or become inaccessible, agroforestry systems help people to diversify production and income and to protect themselves from shortages of fuel and wood.

In Bangladesh, natural forest formations cover less than 6 per cent of the country and the population growth rate is extremely high. An inventory of homestead/village forests in the country indicated that trees outside the forest constitute a vital resource for local populations, providing food, fodder and fuelwood.

The sampling method was based on dual village/household sampling with an agro-ecological and administrative sampling base. Rural Bangladesh was divided into six major regions considered as agro-ecological strata, each subdivided into *thanas* (administrative entities, subdistricts). The households making up the sampling units were chosen at random from a number of villages. The inventory sampled data on palm trees and cane as well as trees, bamboo and thickets.

The results, expressed per stratum and per inhabitant, provide volumetric data for fuelwood and sawnwood and species data for total amounts under and over 20 cm. This inventory was apparently the first to nationwide assessment of trees outside the classified forests in Bangladesh.

METHODS AND TOOLS FOR FUTURE ASSESSMENTS

A priority challenge of future assessments is to know the state and dynamics of all tree resources both in and outside the forest. A country embarking on a planning exercise cannot confine itself solely to the trees within its forests, especially when its wood resources appear to be insufficient.

The choice of tools and methods used to describe or assess trees outside the forest depends on the scale of analysis, kind of data and degree of exactitude desired. The tools used are not generally specific or new; rather, they are combined and implemented in original ways.

The inventory in Bangladesh described above is one of the numerous examples of methods developed for gathering data on TOF. The Bangladesh study gives evidence of the adaptations needed to bring conventional forest inventory procedures in line with the specific nature of this resource.

In some aspects - e.g. structure, spatial distribution and extent of area cover - trees outside the forest are more difficult to assess than forest formations. The assessment of TOF does not lend itself to the potential cost savings associated with expanded uses of remote sensing technology. Remote sensing by satellite presents more difficulties for assessing TOF resources than for assessing attributes such as forest area. However, satellite data do allow a region to be stratified on the basis of ecological criteria and land cover, providing the basis for a good working document for more specific work in the future. The most commonly used remote sensing technology for TOF resources is aerial photography, which can be used to describe spatial distribution and to distinguish TOF cover classifications, providing the appropriate scale is chosen. However, high costs prohibit widespread use of aerial photography for TOF assessments in most countries. The new 1 m resolution satellite sensors represent a possible future alternative to aerial photography.

Some TOF field inventories are modelled on forest inventory methods and keep to biological and physical criteria; others emphasize social aspects, choosing villages as the sampling units. For measurements on the ground, sampling arrangements designed for forest stands may not be the most effective arrangements for trees.

Less traditional sampling plans which would theoretically be better suited to this resource should be tested on various categories of TOF, especially those covering fairly large areas.

Studies of the social and economic benefits or impacts of TOF often rely on household surveys, interviews or standardized appraisals such as rapid or participatory rural appraisal. The integration of the last two approaches - biophysical inventory and socio-economic analysis - is not simple and calls for caution given the great variety of social situations that are only meaningful in the local context.

Environmental benefits or impacts of TOF might be indirectly assessed by linking measurable indicators, such as the number and type of trees, with environmental variables such as water quality or erosion. In an urban setting, tree cover might have direct impact on the ambient temperature. Measuring the environmental impact of tree management is an issue for all natural resource planning or management operations.

Assessment of trees outside the forest requires geographical, ecological, biophysical, social and economic data. However, this implies that an important amount of information will have to be carefully processed. The diversity of end-uses for this information, including land use planning and analysis based on inventory, will need to be considered in data assembly and processing and in the presentation of results.

It is important to know the status of trees outside the forest at any given moment, but it is even more essential to be able to trace patterns of change over time in the same area. The two most commonly used approaches have been comparison of aerial photos taken at sufficiently long intervals and surveys among villagers/managers combined with field inventories.

Some countries, such as France and the United Kingdom, have undertaken periodic inventories based on the establishment of permanent plots linked to permanent forest inventories. However, the high cost of this type of operation limits the number of countries able to adopt it. India and Bangladesh are now experimenting with options for the future.

The current trend towards decentralized authority in land use planning suggests the importance of carrying out assessments at the local level, where the geographical, historical and socio-economic context is relatively harmonious. A minimum number of common rules concerning methods and arrangements is necessary, however, if the data are to be comparable at the country level. Certainly, the technical side of assessing trees outside the forest is complex and more research is needed to better pinpoint the resource.

5

A Forest Ecosystem and Biochemistry

Nutrient Cycling

In ecology, a biogeochemical cycle / nutrient cycle is a circuit or pathway by which a chemical element or molecule moves through both biotic ("bio-") and abiotic ("geo-") compartments of an ecosystem. In effect, the element is recycled, although in some such cycles there may be places (called "sinks") where the element is accumulated or held for a long period of time.

All chemical elements occurring in organisms are part of biogeochemical cycles. In addition to being a part of living organisms, these chemical elements also cycle through abiotic factors of ecosystems such as water (hydrosphere), land (lithosphere), and the air (atmosphere); the living factors of the planet can be referred to collectively as the biosphere. All the chemicals, nutrients, or elements—such as carbon, nitrogen, oxygen, phosphorus—used in ecosystems by living organisms operate on a closed system, which refers to the fact that these chemicals are recycled instead of being lost and replenished constantly such as in an open system. The energy of an ecosystem occurs on an open system; the sun constantly gives the planet energy in the form of light while it is eventually used and lost in the form of heat throughout the trophic levels of a food web.

The Earth does not constantly receive more chemicals as it receives light. The Earth only has those chemicals that were formed in the creation of the Earth, and the only way to obtain more chemicals or nutrients is from occasional meteorites from outer space that contain those elements. Because chemicals operate on a closed system and cannot be lost and replenished like energy can, these chemicals must be recycled throughout all of Earth's processes that use those chemicals or elements. These cycles include both the living biosphere, and the nonliving lithosphere, atmosphere, and hydrosphere. The term "biogeochemical" takes its prefixes from these cycles: Bio refers to the biosphere. Geo refers collectively to the lithosphere, atmosphere, and hydrosphere. Chemical, of course, refers to the chemicals that go through the cycle.

The chemicals are sometimes held for long periods of time in one place. This place is called a reservoir, which, for example, includes such things as coal deposits that are storing carbon for a long period of time. When chemicals are held for only short periods of time, they are being held in exchange pools. Generally, reservoirs are abiotic factors while exchange pools are biotic factors. Examples of exchange pools include plants and animals, which temporarily use carbon in their systems and release it back into the air or wherever. Carbon is held for a relatively short time in plants and animals when compared to coal deposits. The amount of time that a chemical is held in one place is called its residence.

The most well-known and important biogeochemical cycles, for example, include the carbon cycle, the nitrogen cycle, the oxygen cycle, the phosphorus cycle, and the water cycle.

Biogeochemical cycles always involve equilibrium states: a balance in the cycling of the element between compartments. However, overall balance may involve compartments distributed on a global scale.

Biogeochemical cycles of particular interest in ecology are:

- Nitrogen cycle
- Oxygen cycle
- Carbon cycle
- Phosphorus cycle
- Sulphur cycle
- Water cycle
- Hydrogen cycle

Nitrogen Cycle:

The nitrogen cycle is a much more complicated biogeochemical cycle but also cycles through living parts and nonliving parts including the water, land, and air. Nitrogen is a very important molecule in that it is part of both proteins, present in the composition of the amino acids that make up proteins, as well as nucleic acids such as DNA and RNA, present in nitrogenous bases. The largest reservoir of nitrogen is the atmosphere, in which about 78% of nitrogen is contained as nitrogen gas (N2). Nitrogen gas is "fixed," in a process called nitrogen fixation. Nitrogen fixation combines nitrogen with oxygen to create nitrates (NO3).

Nitrates can then be used by plants or animals (which eat plants or eat animals that have eaten plants). Nitrogen can be fixed either by lightning, industrial methods (such as for fertilizer), in free nitrogen-fixing bacteria in the soil, as well as in nitrogen-fixing bacteria present in roots of legumes (such as rhizobium). Nitrogen-fixing bacteria use certain enzymes that are capable of fixing nitrogen gas into nitrates and include free bacteria in soil, symbiotic bacteria in legumes, and also cyanobacteria, or blue-green algae, in water.

After being used by plants and animals, nitrogen is then disposed of in decay and wastes. Detritivores and decomposers decompose the detritius from plants and animals, nitrogen is changed into ammonia, or nitrogen with 3 hydrogen atoms (NH3). Ammonia is toxic and cannot be used by plants or animals, but nitrite bacteria present in the soil can take ammonia and turn it into nitrite, nitrogen with two oxygen atoms (NO2). Although nitrite is also unusable by most plants and animals, nitrate bacteria changes nitrites back into nitrates, usable by plants and animals. Some nitrates are also converted back into nitrogen gas through the process of denitrification, which is the opposite of nitrogen-fixing, also called nitrification. Certain denitrifying bacteria are responsible for this.

oxygen cycle

The oxygen cycle is the biogeochemical cycle that describes the movement of oxygen within and between its three main reservoirs: the atmosphere, the biosphere, and the lithosphere. The main driving factor of the oxygen cycle is photosynthesis, which is responsible for the modern Earth's atmosphere and life as we know it. If all photosynthesis were to cease, the Earth's atmosphere would be devoid of all but trace amounts of oxygen within 5000 years. The oxygen cycle would no longer exist.

carbon cycle

The carbon cycle is the biogeochemical cycle by which carbon is exchanged between the biosphere, geosphere, hydrosphere and atmosphere of the Earth. Other bodies may have carbon cycles, but little is known about them. All of these components are reservoirs of carbon. The cycle is usually thought of as four main reservoirs of carbon interconnected by pathways of exchange. The reservoirs are the atmosphere, terrestrial biosphere (usually includes freshwater systems), oceans, and sediments (includes fossil fuels). The annual movements of carbon, the carbon exchanges between reservoirs, occur because of various chemical, physical, geological, and biological processes. The ocean contains the largest pool of carbon near the surface of the Earth, but most of that pool is not involved with rapid exchange with the atmosphere.

The global carbon budget is the balance of the exchanges (incomes and losses) of carbon between the carbon reservoirs or between one specific loop (e.g., atmosphere - biosphere) of the carbon cycle. An examination of the carbon budget of a pool or reservoir can provide information about whether the pool or reservoir is functioning as a source or sink for carbon dioxide.

phosphorus cycle

The phosphorus cycle is the biogeochemical cycle that describes the movement of phosphorus through the lithosphere, hydrosphere, and biosphere.

Unlike many other biogeochemicals, the atmosphere does not play a significant role in the movements of phosphorus, because phosphorus and phosphorus-based compounds are usually solids at the typical ranges of temperature and pressure found on Earth.

sulphur cycle

Sulphur is one of the constituents of many proteins and vitamins and hormones. It recycles like other biogeochemical cycles.

The essential steps of the sulphur cycle are:

Mineralization of organic sulphur to the inorganic form, hydrogen sulfide: (H_2S).

Oxidation of sulfide and elemental sulphur (S) and related compounds to sulfate (SO_4^{2-}).

Reduction of sulfate to sulfide.

Microbial immobilization of the sulphur compounds and subsequent incorporation into the organic form of sulphur.

water cycle

The water cycle — technically known as the hydrologic cycle — is the continuous circulation of water within the Earth's hydrosphere, and is driven by solar radiation. This includes the atmosphere, land, surface water and groundwater. As water moves through the cycle, it changes state between liquid, solid, and gas phases. Water moves from compartment to compartment, such as from river to ocean, by the physical processes of evaporation, precipitation, infiltration, runoff, and subsurface flow. Movement of water within the water cycle is the subject of the field of hydrology.

hydrogen cycle

Hydrogen is one of the constituents of water. It recycles as in other biogeochemical cycles. It is actively involved with the other cycles like the carbon cycle, nitrogen cycle and sulphur cycle as well.

Anaerobic decomposition of organic substances to carbon dioxide and methane is a collaborative effort involving many different reactions and species of microorganisms. This is also called *interspecies hydrogen transfer*.

Nutrients for plant growth

Plant nutrition often is confused with fertilization. Plant nutrition refers to a plant's need for and use of basic chemical elements. Fertilization is the term used when these materials are added to the environment around a plant. A lot must happen before a chemical element in a fertilizer can be used by a plant. Plants need 17 elements for normal growth. Three of them—carbon, hydrogen, and oxygen—are found in air and water. The rest are found in the soil.

Six soil elements are called macronutrients because they are used in relatively large amounts by plants. They are nitrogen, potassium, magnesium, calcium, phosphorus, and sulphur in Table.

Table . Plant macronutrients.

Element	Absorbed as	Leaches from soil/ Mobility in plant	Signs of excess	Signs of deficiency	Notes
Nitrogen (N)	NO_3-, (nitrate), NH_4+ (ammonium)	Leachable, especially NO_3-. Mobile in plants.	Succulent growth; dark green color; weak, spindly growth; few fruits. May cause brittle growth, especially under high temperatures.	Reduced growth, yellowing (chlorosis). Reds and purples may intensify in some plants. Reduced lateral bud breaks. Symptoms appear first on older growth.	In general, the best NH_4+:NO_3- ratio is 1:1. Under low sugar conditions (low light), high NH_4+ can cause leaf curl. Uptake is inhibited by high P levels. The N:K ratio is extremely important. Indoors, the best N:K ratio is 1:1 unless light is extremely high. In soils with a high C:N ratio, more N should be supplied.
Phosphorus (P)	H_2PO_4-, HPO_4- (phosphate)	Normally not leachable, but may leach from soil high in bark or peat. Not readily mobile in plants.	Shows up as micronutrient deficiency of Zn, Fe, or Co.	Reduced growth. Color may intensify; browning or purpling of foliage in some plants. Thin stems, reduced lateral bud breaks, loss of lower leaves, reduced flowering.	Rapidly bound (fixed) on soil (P) particles. Under acid conditions, fixed with Fe, Mg, and Al. Under alkaline conditions, fixed with Ca. Important for young plant and seedling growth. High P interferes with micronutrient absorption and N absorption. Used in relatively small amounts when compared to N and K.
Potassium (K)	K+	Can leach in sandy soils. Mobile in plants.	Causes N deficiency in plant and may affect the uptake of other positive ions.	Reduced growth, shortened internodes. Marginal burn or scorch (brown leaf edges), necrotic (dead) spots in leaves. Reduction of lateral bud breaks, tendency to wilt readily.	N:K balance is important. High N:low K favors vegetative growth; low N:high K promotes reproductive growth (flowers, fruit).

Eight other soil elements are used in much smaller amounts and are called micronutrients or trace elements in Table.

Table. Plant micronutrients.

Element	Absorbed as	Signs of excess	Signs of deficiency	Notes
Iron (Fe)	Fe++, Fe+++	Rare except on flooded soils. Interveinal chlorosis, primarily on young tissue, which eventually may turn white.	Soil high in Ca, Mn, P, or heavy metals (Cu, Zn); high pH; poorly drained soil; oxygen-deficient soil; nematode attack on roots.	Add Fe in the chelate form. The type of chelate needed depends on soil pH.
Boron (B)	BO3- Borate	Blackening or death of tissue between veins.	Failure to set seed, internal breakdown, death of apical buds.	
Zinc (Zn)	Zn++	Shows up as Fe deficiency. Also interferes with Mg absorption.	"Little leaf" (reduction in leaf size), short internodes, distorted or puckered leaf margins, interveinal chlorosis.	
Copper (Cu)	Cu++, Cu+	Can occur at low pH. Shows up as Fe deficiency.	New growth small, mis-shapen, wilted.	May be found in some peat soils.
Manganese (Mn)	Mn++	Reduction in growth, brown spotting on leaves. Shows up as Fe deficiency.	Interveinal chlorosis of leaves followed by brown spots, producing a checkered effect.	Found under acid conditions.
Molybdenum (Mo)	MoO4- (molybdate)		Interveinal chlorosis on older or midstem leaves, twisted leaves (whiptail).	
Chlorine (Cl)	Cl-	Salt injury, leaf burn. May increase succulence.	Leaves wilt, then become bronze, then chlorotic, then die; club roots.	

They are iron, zinc, molybdenum, manganese, boron, copper, cobalt, and chlorine.

Most of the nutrients a plant needs are dissolved in water and then absorbed by its roots. In fact, 98 percent are absorbed from the soil-water solution, and only about 2 percent are actually extracted from soil particles.

Fertilizers

Fertilizers are materials containing plant nutrients that are added to the environment around a plant. Generally, they are added to the water or soil, but some can be sprayed on leaves. This method is called foliar fertilization. It should be done carefully with a dilute solution, because a high fertilizer concentration can injure leaf cells. The nutrient, however, does need to pass through the thin layer of wax (cutin) on the leaf surface.

Fertilizers are not plant food! Plants produce their own food from water, carbon dioxide, and solar energy through photosynthesis. This food (sugars and carbohydrates) is combined with plant nutrients to produce proteins, enzymes, vitamins, and other elements essential to growth.

Nutrient absorption

Anything that reduces or stops sugar production in leaves can lower nutrient absorption. Thus, if a plant is under stress because of low light or extreme temperatures, nutrient deficiency may develop.

A plant's developmental stage or rate of growth also may affect the amount of nutrients absorbed. Many plants have a rest (dormant) period during part of the year. During this time, few nutrients are absorbed. Plants also may absorb different nutrients as flower buds begin to develop than they do during periods of rapid vegetative growth.

A nutrient is any element or compound necessary for or contributing to an organism's metabolism, growth, or other functioning. Six nutrient groups exist, classifiable as those that provide energy, and as those that otherwise support metabolic processes in the body: Some of them are essential because they cannot be synthesized in the body and must be obtained from a food source.

Substances that provide energy

Carbohydrates are compounds made up of sugars. Carbohydrates are classified by their number of sugar units: monosaccharides (such as glucose and fructose), disaccharides (such as sucrose and lactose), oligosaccharides, and polysaccharides (such as starch, glycogen, and cellulose).

Proteins are organic compounds that consists of amino acids joined by peptide bonds. The body does not manufacture certain amino acids (termed essential amino acids); the diet must supply these. In nutrition, proteins are broken down through digestion back into free amino acids.

Fats consist of a glycerin molecule with three fatty acids attached. Fatty acids are unbranched hydrocarbon chains, connected by single bonds alone (saturated fatty acids) or by both single and double bonds (unsaturated fatty acids). Fats are needed to keep cells functioning properly, to insulate body organs against shock, to keep body temperature stable, and to maintain healthy skin and hair. The body does not manufacture certain fatty acids (termed essential fatty acids); the diet must supply these.

Fat has an energy content of 9 kcal/g (~37.7 kJ/g); proteins and carbohydrates 4 kcal/g (~16.7 kJ/g). Ethanol (grain alcohol) has an energy content of 7 kcal/g (~29.3 kJ/g).

Substances that support metabolism

Minerals are generally trace elements, salts, or ions such as copper and iron. These minerals are essential to human metabolism.

Vitamins are organic compounds essential to the body. They usually act as coenzymes for various proteins in the body.

Water is an essential nutrient and is directly involved in all the chemical reactions of life.[citation needed]

Nutrition as a science

Any classification of "nutrients" is likely to be arbitrary given the status of nutrition as a developing science. Researchers are becoming more aware of a wider range of nutrients essential for health.

An organism will metabolise any organic compound to use for its energy content, for structural purposes (growth or replacement of living structures), or for participation in chemical reactions necessary for life. Any particular substance can play more than one role in the body, though researchers lack a good understanding of these roles.

The discovery of the group of nutrients called phytonutrients reinforces the provisional nature of our knowledge. We know little about phytonutrients, organic compounds from plants, which play an essential role in the normal functioning of a body and have complex hormonal effects on health, or play an active role in the amelioration of disease. They do not fit readily into the scheme of the traditional nutrition categories.

Nutrients and the environment

The band of a green alga (*Enteromorpha*) along this shore indicates that there is a nearby source of nutrients (probably nitrates or ammonia from a small estuary).

While in essence true to the definition above, the term nutrients has a more limited meaning within the specialised fields of water quality and water pollution, referring specifically to plant fertilizers. In this context, certain mineral compounds can have an adverse impact on water quality because of their ability to promote plant and algae growth. An excessive growth of aquatic plants can clog waterways , and over-stimulation of algae and microbes leads to an ecological process called eutrophication.

A surprisingly small number of elements provide interest or concern in this context: really just nitrogen and phosphorus in most aquatic systems. Mineral compounds involved are ammonia, nitrites, nitrates, and orthophosphates. Organic compounds also may contribute, in as much as they also contain nitrogen and phosphorus. The reason only a few chemicals are of concern has to do with the fact that plants are made up mostly of compounds of carbon (C), hydrogen (H), oxygen (O), nitrogen (N), and phosphorus (P), and lesser amounts of sulphur (S), potassium (K), magnesium (Mg), and calcium (Ca). These elements constitute the macronutrients. Many other elements, though necessary for growth, classify as micronutrients due to the very small quantities required.

Plants obtain carbon, hydrogen, and oxygen (elements most needed for growth) from the air and water, where all three elements occur in great abundance as water and as carbon dioxide.

Nutrients having greatest potential to influence plant growth in aquatic environments is those elements needed for plant growth in proportionately large amounts (that is, macronutrients) but likely to become limiting—that is, present in amounts that could be depleted by continued growth. Once used up, further growth will not be possible. Of the nine macronutrients, nitrogen and phosphorus are most likely to become limiting. The others always remain present in great abundance (C, H, O) or usually in amounts that exceed the requirements of aquatic plants or algae.

Farmers apply fertilizer nutrients in the form of nitrogen, phosphorus, and potassium (N, P, and K with perhaps micronutrients) to prevent these elements from becoming limiting in the soil. These elements become concentrated in wastewaters from animal pens and septic or sewage systems. And these elements (especially N and P) in runoff or wastewater discharges reaching streams, lakes, or seas will promote aquatic plant growth. Abundant plant growth itself gives cause for concern in assessing water quality. The most abundant "plants" in most aquatic environments are algae. When essential nutrients are plentiful, algae multiply. If these algae are microscopic phytoplankton, their growth increases the turbidity of the water. The water then becomes cloudy, colored a shade of green, yellow, or brown (sometimes red). A super abundance of algae, or of higher plants, in an aquatic system can signal excessive inputs of nutrients.

The pathway of nitrogen

Role of nitrogen in the biosphere

The growth of all organisms depends on the availability of mineral nutrients, and none is more important than nitrogen, which is required in large amounts as an essential component of proteins, nucleic acids and other cellular constituents. There is an abundant supply of nitrogen in the earth's atmosphere - nearly 79% in the form of N2 gas. However, N2 is unavailable for use by most organisms because there is a triple bond between the two nitrogen atoms, making the molecule almost inert. In order for nitrogen to be used for growth it must be "fixed" (combined) in the form of ammonium (NH4) or nitrate (NO3) ions. The weathering of rocks releases these ions so slowly that it has a neglible effect on the availability of fixed nitrogen. So, nitrogen is often the limiting factor for growth and biomass production in all environments where there is suitable climate and availability of water to support life.

Microorganisms have a central role in almost all aspects of nitrogen availability and thus for life support on earth:

Some bacteria can convert N2 into ammonia by the process termed nitrogen fixation; these bacteria are either free-living or form symbiotic associations with plants or other organisms (e.g. termites, protozoa) other bacteria bring about transformations of ammonia to nitrate, and of nitrate to N2 or other nitrogen gases many bacteria and fungi degrade organic matter, releasing fixed nitrogen for reuse by other organisms.

All these processes contribute to the nitrogen cycle.

We shall deal first with the process of nitrogen fixation and the nitrogen-fixing organisms, then consider the microbial processes involved in the cycling of nitrogen in the biosphere.

Nitrogen fixation

A relatively small amount of ammonia is produced by lightning. Some ammonia also is produced industrially by the Haber-Bosch process, using an iron-based catalyst, very high pressures and fairly high temperature. But the major conversion of N2 into ammonia, and thence into proteins, is achieved by microorganisms in the process called nitrogen fixation (or dinitrogen fixation).

The table below shows some estimates of the amount of nitrogen fixed on a global scale. The total biological nitrogen fixation is estimated to be twice as much as the total nitrogen fixation by non-biological processes.

Type of fixation N2 fixed (1012 g per year, or 106 metric tons per year)

Type of fixation	N_2 fixed (10^{12} g per year, or 10^6 metric tons per year)
Non-biological	
Industrial	about 50
Combustion	about 20
Lightning	about 10
Total	**about 80**
Biological	
Agricultural land	about 90
Forest and non-agricultural land	about 50
Sea	about 35
Total	**about 175**

To illustrate the importance of biological nitrogen fixation, the image below shows part of the Lower Sonoran desert in Arizona. Every plant in this scene depends ultimately on biological nitrogen fixation. Both free-living cyanobacteria and the cyanobacterial associates of lichens initially contributed nitrogen to the soil by forming a cryptobiotic crust. Now numerous leguminous plants occur in this desert, with nitrogen-fixing Rhizobium in their root nodules. Examples are the green-stemmed brush-like trees at the right and left of the image

(Parkinsonia species, common name "paloverde"), and several acacias and mesquites .

Mechanism of biological nitrogen fixation

Biological nitrogen fixation can be represented by the following equation, in which two moles of ammonia are produced from one mole of nitrogen gas, at the expense of 16 moles of ATP and a supply of electrons and protons (hydrogen ions):

$$N_2 + 8H+ + 8e- + 16\ ATP = 2NH_3 + H_2 + 16ADP + 16\ Pi$$

This reaction is performed exclusively by prokaryotes (the bacteria and related organisms), using an enzyme complex termed nitrogenase. This enzyme consists of two proteins - an iron protein and a molybdenum-iron protein, as shown below.

The reactions occur while N2 is bound to the nitrogenase enzyme complex. The Fe protein is first reduced by electrons donated by ferredoxin. Then the reduced Fe protein binds ATP and reduces the molybdenum-iron protein, which donates electrons to N2, producing HN=NH. In two further cycles of this process (each requiring electrons donated by ferredoxin) HN=NH is reduced to H2N-NH2, and this in turn is reduced to 2NH3.

Depending on the type of microorganism, the reduced ferredoxin which supplies electrons for this process is generated by photosynthesis, respiration or fermentation.

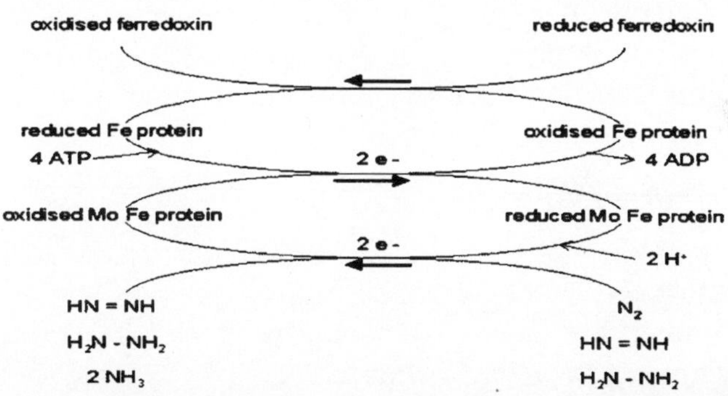

There is a remarkable degree of functional conservation between the nitrogenase proteins of all nitrogen-fixing bacteria. The Fe protein and the Mo-Fe protein have been isolated from many of these bacteria, and nitrogen fixation can be shown to occur in cell-free systems in a laboratory when the Fe protein of one species is mixed with the Mo-Fe protein of another bacterium, even if the species are very distantly related.

The nitrogen-fixing organisms

All the nitrogen-fixing organisms are prokaryotes (bacteria). Some of them live independently of other organisms - the so-called free-living nitrogen-fixing bacteria. Others live in intimate symbiotic associations with plants or with other organisms (e.g. protozoa). Examples are shown in the table below.

A point of special interest is that the nitrogenase enzyme complex is highly sensitive to oxygen. It is inactivated when exposed to oxygen, because this reacts with the iron component of the proteins. Although this is not a problem for anaerobic bacteria, it could be a major problem for the aerobic species such as cyanobacteria (which generate oxygen during photosynthesis) and the free-living aerobic bacteria of soils, such as Azotobacter and Beijerinckia. These organisms have various methods to overcome the problem. For example, Azotobacter species have the highest known rate of respiratory metabolism of any organism, so they might protect the enzyme by maintaining a very low level of oxygen in their cells. Azotobacter species also produce copious amounts of extracellular polysaccharide (as do Rhizobium species in culture -). By maintaining water within the polysaccharide slime layer, these bacteria can limit the diffusion rate of oxygen to the cells. In the symbiotic nitrogen-fixing organisms such as Rhizobium, the root nodules can contain oxygen-scavenging molecules such as leghaemoglobin, which shows as a pink colour when the active nitrogen-fixing nodules of legume roots are cut open. Leghaemoglobin may regulate the supply of oxygen to the nodule tissues in the same way as haemoglobin regulates the supply of oxygen to mammalian tissues. Some of the cyanobacteria have yet another mechanism for protecting nitrogenase: nitrogen fixation occurs in special cells (heterocysts) which possess only photosystem I (used to generate ATP by light-mediated reactions) whereas the other cells have both photosystem I and photosystem II (which generates oxygen when light energy is used to split water to supply H2 for synthesis of organic compounds) .

Symbiotic nitrogen fixation

The most familiar examples of nitrogen-fixing symbioses are the root nodules of legumes (peas, beans, clover, etc.).

Part of a clover root system bearing naturally occurring nodules of Rhizobium. Each nodule is about 2-3 mm long.

Clover root nodules at higher magnification, showing two partly crushed nodules (arrowheads) with pink-coloured contents. This colour is caused by the presence of the pigment leghaemoglobin - a unique metabolite of this type of symbiosis. Leghaemoglobin is found only in the nodules and is not produced by either the bacterium or the plant when grown alone.

In these leguminous associations the bacteria usually are Rhizobium species, but the root nodules of soybeans, chickpea and some other legumes are formed by small-celled rhizobia termed Bradyrhizobium. Nodules on some tropical leguminous plants are formed by yet other genera. In all cases the bacteria "invade" the plant and cause the formation of a nodule by inducing localised proliferation of the plant host cells. Yet the bacteria always remain separated from the host cytoplasm by being enclosed in a membrane - a necessary feature in symbioses .

Part of a crushed root nodule of a pea plant, showing four root cells containing colonies of Rhizobium. The nuclei (n) of two root cells are shown; cw indicates the cell wall that separates two plant cells. Although it cannot be seen clearly in this image, the bacteria occur in clusters which are enclosed in membranes, separating them from the cytoplasm of the plant cells.

In nodules where nitrogen-fixation is occurring, the plant tissues contain the oxygen-scavenging molecule, leghaemoglobin (serving the same function as the oxygen-carrying haemoglobin in blood). The function of this molecule in nodules is to reduce the amount of free oxygen, and thereby to protect the nitrogen-fixing enzyme nitrogenase, which is irreversibly inactivated by oxygen.

Some excellent images and discussion of these leguminous associations can be found at:

Frankia is a genus of the bacterial group termed actinomycetes - filamentous bacteria that are noted for their production of air-borne spores. Included in this group are the common soil-dwelling Streptomyces species which produce many of the antibiotics used in medicine .

Frankia species are slow-growing in culture, and require specialised media, suggesting that they are specialised symbionts. They form nitrogen-fixing root nodules (sometimes called actinorhizae) with several woody plants of different families, such as alder (Alnus species), sea buckthorn (Hippophae rhamnoides, which is common in sand-dune environments) and Casuarina (a Mediterranean tree genus).

Alder and the other woody hosts of Frankia are typical pioneer species that invade nutrient-poor soils. These plants probably benefit from the nitrogen-fixing association, while supplying the bacterial symbiont with photosynthetic products.

The photosynthetic cyanobacteria often live as free-living organisms in pioneer habitats such as desert soils or as symbionts with lichens in other pioneer habitats. They also form symbiotic associations with other organisms such as the water fern Azolla, and cycads.The association with Azolla, where cyanobacteria (Anabaena azollae) are harboured in the leaves, has sometimes been shown to be important for nitrogen inputs in rice paddies, especially if the fern is allowed to grow and then ploughed into the soil to release nitrogen before the rice crop is sown.

A symbiotic association of cyanobacteria with cycads is shown below. The first image shows a pot-grown plant. The second image shows a close-up of the soil surface in this pot. Short, club-shaped, branching roots have grown into the aerial environment. These aerial roots contain a nitrogen-fixing cyanobacterial symbiont.

In addition to these intimate and specialised symbiotic associations, there are several free-living nitrogen-fixing bacteria that grow in close association with plants. For example, Azospirillum species have been shown to fix nitrogen when growing in the root zone (rhizosphere) or tropical grasses, and even of maize plants in field conditions. Similarly, Azotobacter species can fix nitrogen in the rhizosphere of several plants. In both cases the bacteria grow at the expense of sugars and other nutrients that leak from the roots. However, these bacteria can make only a small contribution to the nitrogen nutrition of the plant, because nitrogen-fixation is an energy-expensive process, and large amounts of organic nutrients are not continuously available to microbes in the rhizosphere.

1. Uptake of NH_4 or NO_3 by organisms
2. Release of NH_4 by decomposition
3,4. Microbial oxidation of NH_4 (yields energy in aerobic conditions)
5. Denitrification (NO_3 respiration) by microbes in anaerobic conditions (NO_3 is used instead of O_2 as the terminal electron acceptor during decomposition of organic matter)
6. Nitrogen fixation
7. Nitrate leaching from soil

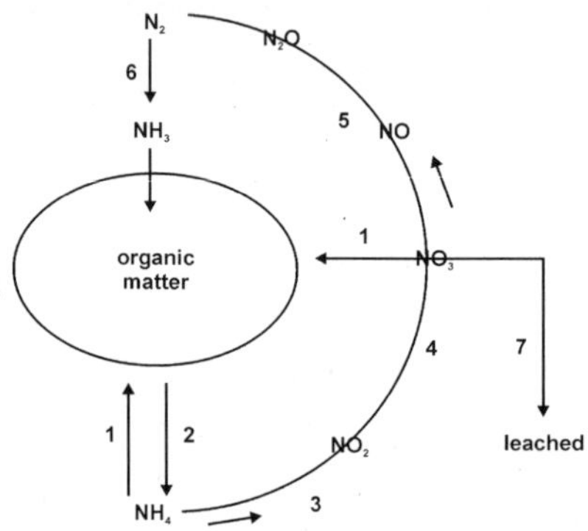

This limitation may not apply to the bacteria that live in root nodules or other intimate symbiotic associations with plants. It has been estimated that nitrogen fixation in the nodules of clover roots or other leguminous plants may consume as much as 20% of the total photosynthate.

The nitrogen cycle

At any one time a large proportion of the total fixed nitrogen will be locked up in the biomass or in the dead remains of organisms. So, the only nitrogen available to support new growth will be that which is supplied by nitrogen fixation from the atmosphere or by the release of ammonium or simple organic nitrogen compounds through the decomposition of organic matter.

Nitrification

The term nitrification refers to the conversion of ammonium to nitrate (pathway 3-4). This is brought about by the nitrifying bacteria, which are specialised to gain their energy by oxidising ammonium, while using CO2 as their source of carbon to synthesise organic compounds. Organisms of this sort are termed chemoautotrophs - they gain their energy by chemical oxidations (chemo-) and they are autotrophs (self-feeders) because they do not depend on pre-formed organic matter. In principle the oxidation of ammonium by these bacteria is no different from the way in which humans gain energy by oxidising sugars. Their use of CO_2 to produce organic matter is no different in principle from the behaviour of plants.

The nitrifying bacteria are found in most soils and waters of moderate pH, but are not active in highly acidic soils. They almost always are found as mixed-species communities (termed consortia) because some of them - e.g. Nitrosomonas species - are specialised to convert ammonium to nitrite (NO_2-) while others - e.g. Nitrobacter species - convert nitrite to nitrate (NO_3-). In fact, the accumulation of nitrite inhibits Nitrosomonas, so it depends on Nitrobacter to convert this to nitrate, whereas Nitrobacter depends on Nitrosomonas to generate nitrite.

The nitrifying bacteria have some important environmental consequences, because they are so common that most of the ammonium in oxygenated soil or natural waters is readily converted to nitrate. Most plants and microorganisms can take up either nitrate or ammonium. However, process of nitrification has some undesirable consequences.

The ammonium ion (NH_4+) has a positive charge and so is readily adsorbed onto the negatively charged clay colloids and soil organic matter, preventing it from being washed out of the soil by rainfall. In contrast, the negatively charged nitrate ion is not held on soil particles and so can be washed down the soil profile - the process termed leaching. In this way, valuable nitrogen can be lost from the soil, reducing the soil fertility. The nitrates can then accumulate in groundwater, and ultimately in drinking water. There are strict regulations governing the amount of nitrate that can be present in drinking water, because nitrates can be reduced to highly reactive nitrites by microorganisms in the anaerobic conditions of the gut. Nitrites are absorbed from the gut and bind to haemoglobin, reducing its oxygen-carrying capacity. In young babies this can lead to respiratory distress - the condition known as "blue baby syndrome". Nitrite in the gut also can react with amino compounds, forming highly carcinogenic nitrosamines.

Denitrification

Denitrification refers to the process in which nitrate is converted to gaseous compounds (nitric oxide, nitrous oxide and N_2) by microorganisms. The sequence usually involves the production of nitrite (NO_2-) as an intermediate step.Several types of bacteria perform this conversion when growing on organic matter in anaerobic conditions. Because of the lack of oxygen for normal aerobic respiration, they use nitrate in place of oxygen as the terminal electron acceptor. This is termed anaerobic respiration and can be illustrated as follows:

In aerobic respiration (as in humans), organic molecules are oxidised to obtain energy, while oxygen is reduced to water:

$$C_6H_{12}O_6 + 6\ O_2 = 6\ CO_2 + 6\ H_2O + \text{energy}$$

In the absence of oxygen, any reducible substance such as nitrate (NO_3-) could serve the same role and be reduced to nitrite, nitric oxide, nitrous oxide or N_2.

Thus, the conditions in which we find denitrifying organisms are characterised by (1) a supply of oxidisable organic matter, and (2) absence of oxygen but availability of reducible nitrogen sources. A mixture of gaseous nitrogen products is often produced because of the stepwise use of nitrate, nitrite, nitric oxide and nitrous oxide as electron acceptors in anaerobic respiration.

The common denitrifying bacteria include several species of Pseudomonas, Alkaligenes and Bacillus. Their activities result in substantial losses of nitrogen into the atmosphere, roughly balancing the amount of nitrogen fixation that occurs each year.

Phosphorus and sulphur

Phosphorus

Phosphorus, (from the Greek language *phôs* meaning "light", and *phoros* meaning "bearer"), is the chemical element in the periodic table that has the symbol P and atomic number 15. A multivalent nonmetal of the nitrogen group, phosphorus is commonly found in inorganic phosphate rocks and in all living cells. Due to its high reactivity, it is never found as a free element in nature. It emits a faint glow upon exposure to oxygen, occurs in several allotropic forms, and is an essential element for living organisms. The most important commercial use of phosphorus is in the production of fertilizers. It is also widely used in explosives, nerve agents, friction matches, fireworks, pesticides, toothpaste, and detergents.

Phosphorus (P), the 15th element on the periodic table with an atomic weight of 30.974, is an essential nutrient for all life forms. Phosphorus plays a role in deoxyribonucleic acid (DNA), ribonucleic acid (RNA), adenosine diphosphate (ADP), and adenosine triphosphate (ATP). Phosphorus is required for these necessary components of life to occur.

Phosphate is usually not readily available for uptake in soils. Phosphate is only freely soluble in acid solutions and under reducing conditions. In the soil it is rapidly immobilized as calcium or iron phosphates. Most of the phosphorus in soils is adsorbed to soil particles or incorporated into organic matter.

Phosphorus in freshwater and marine systems exists in either a particulate phase or a dissolved phase. Particulate matter includes living and dead plankton, precipitates of phosphorus, phosphorus adsorbed to particulates, and amorphous phosphorus. The dissolved phase includes inorganic phosphorus (generally in the soluble orthophosphate form), organic phosphorus excreted by organisms, and macromolecular colloidal phosphorus.

The organic and inorganic particulate and soluble forms of phosphorus undergo continuous transformations. The dissolved phosphorus (usually as orthophosphate) is assimilated by phytoplankton and altered to organic phosphorus. The phytoplankton are then ingested by detritivores or zooplankton. Over half of the organic phosphorus taken up by zooplankton is excreted as inorganic P. Continuing the cycle, the inorganic P is rapidly assimilated by phytoplankton.

Lakes and reservoir sediments serve as phosphorus sinks. Phosphorus-containing particles settle to the substrate and are rapidly covered by sediment. Continuous accumulation of sediment will leave some phosphorus too deep within the substrate to be reintroduced to the water column. Thus, some phosphorus is removed permanently from biocirculation .

Health Effects:

Phosphate: Phosphate itself does not have notable adverse health effects. However, phosphate levels greater than 1.0 may interfere with coagulation in water treatment plants. As a result, organic particles that harbor microorganisms may not be completely removed before distribution.

Environmental Effects:

The growth of macrophytes and phytoplankton is stimulated principally by nutrients such as phosphorus and nitrogen. Nutrient-stimulated primary production is of most concern in lakes and estuaries, because primary production in flowing water is thought to be controlled by physical factors, such as light penetration, timing of flow, and type of substrate available, instead of by nutrients.

Freshwater system impacts: Generally, phosphorus (as orthophosphate) is the limiting nutrient in freshwater aquatic systems. That is, if all phosphorus is used, plant growth will cease, no matter how much nitrogen is available. The natural background levels of total phosphorus are generally less than 0.03 mg/l. The natural levels of orthophosphate usually range from 0.005 to 0.05 mg/l Many bodies of freshwater are currently experiencing influxes of phosphorus and nitrogen from outside sources. The increasing concentration of available phosphorus allows plants to assimilate more nitrogen before the phosphorus is depleted. Thus, if sufficient phosphorus is available, elevated concentrations of nitrates will lead to algal blooms. Although levels of 0.08 to 0.10 mg/l orthophosphate may trigger periodic blooms, long-term eutrophication will usually be prevented if total phosphorus levels and orthophosphate levels are below 0.5 mg/l and 0.05 mg/l, respectively.

Estuarine system impacts: In contrast to freshwater, nitrogen is generally the primary limiting nutrient in the seaward portions of estuarine systems. Here, nitrogen levels control the rate of primary production. If the system is supplied with high levels of nitrogen, algal blooms will occur. Systems may be phosphorus limited, however, or become so when nitrogen concentrations are high and N:P>16:1. Freshwater and estuarine systems: Nutrient-induced production of aquatic plants in both freshwater and estuaries has several detrimental consequences: Extensive growth of rooted aquatic macrophytes will interfere with navigation, aeration, and channel capacity.

Dead macrophytes and phytoplankton settle to the bottom of a water body, stimulating microbial breakdown processes that require oxygen. Eventually, oxygen will be depleted.

Sources:

Nonpoint sources:

Natural: Phosphate deposits and phosphate-rich rocks release phosphorus during weathering, erosion, and leaching. Phosphorus may be released from lake and reservoir bottom sediments during seasonal overturns.

Anthropogenic: The primary anthropogenic nonpoint sources of phosphorus include runoff from 1) land areas being mined for phosphate deposits, 2) agricultural areas, and 3) urban/residential areas. Because phosphorus has a strong affinity for soil, little dissolved phosphorus will be transported in runoff. Instead, the eroded sediments from mining and agricultural areas carry the adsorbed phosphorus to the water body. An additional source is the overboard discharge of phosphorus-containing sewage by boats.

Point sources: Sewage treatment plants provide most of the available phosphorus to surface water bodies. A normal adult excretes 1.3 - 1.5 g of phosphorus per day. Additional phosphorus originates from the use of industrial products, such as toothpaste, detergents, pharmaceuticals, and food-treating compounds. Primary treatment removes only 10% of the phosphorus in the waste stream; secondary treatment removes only 30%. The remainder is discharged to the water body (Smith, 1990). Tertiary treatment is required to remove additional phosphorus from the water. The amount of additional phosphorus that can be removed varies with the success of the treatment technologies used. Available technologies include biological removal and chemical precipitation .

Mode of Transport:

Phosphates are primarily discharged directly into the water body by sewage treatment plants. Phosphorus that is adsorped to sediment particles may be transported in overland flow .

Analytical techniques:

A. Total Phosphorus and Orthophosphate: Analysis involves two procedural steps: 1) conversion of the phosphorus form into dissolved orthophosphate by a digestion method, and 2) colorimetric evaluation of the dissolved orthophosphate concentration.

Step 1: Digestion methods

Perchloric Acid Digestion: Recommended only for extremely difficult-to-analyze samples, such as sediments.

Nitric Acid-Sulphuric Acid Method Recommended for most samples.

Persulfate Oxidation Method This simple method should be cross-checked with one or more thorough techniques and adopted if results are identical.

Step 2: Colorimetric methods

Ascorbic Acid Method: Ammonium molybdate and potassium antimonyl tartrate react with orthophosphate to form a heteropoly acid that is reduced to molybdenum blue by ascorbic acid.

Detection limits: Ranges change with light path used.

Range (mg/l P)	Path (cm)
0.3 - 2.0	0.5
0.15 - 1.3	1.0
0.01 - 0.25	5.0

Interferences: Arsenates react with the molybdate to form a similar blue color. Nitrite and hexavalent chromium interfere to yield results 3% less than actual at 1 mg/l and 10% to 15% less than actual at 10 mg/l.

Automated Ascorbic Acid Reduction Method: Ammonium molybdate and potassium antimonyl tartrate react with orthophosphate in an acid medium to form an antimony- phosphomolybdate complex that forms a blue color suitable for photometric measurements when reduced by ascorbic acid.

Detection limits: 0.001 to 10.0 mg/l P when photometric measurements are performed at 650 to 600 in a 15mm tubular flow cell, or 880 nm in a 50mm tubular flow cell.

Interferences: >50 mg/l Fe(3+), 10 mg/l Cu, and 10 mg/l SiO2. Turbidity, color may interfere. Arsenate provides a positive interference.

Notable characteristics

Common phosphorus forms a waxy white solid that has a characteristic, disagreeable smell similar to that of garlic. Pure forms of the element are colorless and transparent. This nonmetal is not soluble in water, but *is* soluble in carbon disulfide. Pure phosphorus ignites spontaneously in air, burning to produce phosphorus pentoxide.

Forms

Phosphorus is known to exist in three allotropic forms: white, red, and black (other allotropic forms are theoretically possible, but are not known). The most common are red and white phosphorus, both of which consist of networks of tetrahedrally arranged groups of four phosphorus atoms. The tetrahedra of white phosphorus form separate groups; the tetrahedra of red phosphorus are linked into chains. White phosphorus burns on contact with air and on exposure to heat or light.

Phosphorus also exists in kinetically and thermodynamically favored forms. They are separated by a transition temperature of -3.8 °C. One is known as the "alpha" form, the other "beta". Red phosphorus is comparatively stable and sublimates at a vapor pressure of 1 atm at 170 °C but burns from impact or frictional heating. A black phosphorus allotrope exists which has a structure similar to graphite – the atoms are arranged in hexagonal sheet layers and will conduct electricity.

Glow

The glow from phosphorus was the attraction of its discovery around 1669, but the mechanism for that glow was not fully described until 1974.

In 1974 the glow was explained by R. J. van Zee and A. U. Khan. A reaction with oxygen takes place at the surface of the solid (or liquid) phosphorus, forming short-lived molecules HPO and P_2O_2 and they both emit visible light. The reaction is slow and only very little of the intermediates is required to produce the luminescence, hence the extended time the glow continues in a stoppered jar.

Although the term phosphorescence is derived from phosphorus, the reaction is properly called luminescence (glowing by its own reaction, in this case chemoluminescence), not phosphorescence (re-emitting light that previously fell on it).

Applications

Concentrated phosphoric acids, which can consist of 70% to 75% P_2O_5 are very important to agriculture and farm production in the form of fertilizers. Global demand for fertilizers led to large increases in phosphate (PO_4^{3-}) production in the second half of the 20th century. Other uses;

Phosphates are utilized in the making of special glasses that are used for sodium lamps.

Phosphorus is widely used to make organophosphorus compounds, through the intermediates phosphorus chlorides and the two phosphorus sulfides: phosphorus pentasulfide, and phosphorus sesquisulfide. Organophosphorus compounds have many applications, including in plasticizers, flame retardants, pesticides, extraction agents, and water treatment.

Phosphorus sesquisulfide is used in heads of strike-anywhere matches.

This element is also an important component in steel production, in the making of phosphor bronze, and in many other related products.

White phosphorus is used in military applications as incendiary bombs, for smoke-screening as smoke pots and smoke bombs, and in tracer ammunition.

Red phosphorus is essential for manufacturing matchbook strikers, flares, and, most notoriously, methamphetamine.

In trace amounts, phosphorus is used as a dopant for N-type semiconductors. ^{32}P and ^{33}P are used as radioactive tracers in biochemical laboratories .

Biological role

Phosphorus is a key element in all known forms of life. Inorganic phosphorus in the form of the phosphate PO_4^{3-} plays a major role in biological molecules such as DNA and RNA where it forms part of the structural backbone of these molecules. Living cells also utilize phosphate to transport cellular energy via adenosine triphosphate (ATP). Nearly every cellular process that uses energy gets it in the form of ATP. ATP is also important for phosphorylation, a key regulatory event in cells. Phospholipids are the main structural components of all cellular membranes. Calcium phosphate salts are used by animals to stiffen their bones.

An average person contains a little less than 1 kg of phosphorus, about three quarters of which is present in bones and teeth in the form of apatite. A well-fed adult in the industrialized world consumes and excretes about 1-3 g of phosphorus per day in the form of phosphate. Phosphorus is an essential mineral macronutrient, which is studied extensively in soil conservation in order to understand plant uptake from soil systems.

In ecological terms, phosphorus is often a limiting nutrient in many environments, i.e. the availability of phosphorus governs the rate of growth of many organisms. In ecosystems an excess of phosphorus can be problematic, especially in aquatic systems.

History

Phosphorus (Greek *phosphoros* was the ancient name for the planet Venus) was discovered by German alchemist Hennig Brand in 1669 through a preparation from urine. Working in Hamburg, Brand attempted to distill salts by evaporating urine, and in the process produced a white material that glowed in the dark and burned brilliantly. Since that time, phosphorescence has been used to describe substances that shine in the dark without burning.

Phosphorus was first made commercially, for the match industry, in the 19th century, by distilling off phosphorus vapour from precipitated phosphates heated in a retort. The precipitated phosphates made from ground-up bones, that had been de-greased and treated with strong acids. This process became obsolete in the late 1890s when the Electric arc furnace was adapted to reduce phosphate rock.

Early matches used white phosphorus in their composition, which was dangerous due to its toxicity. Murders, suicides and accidental poisonings resulted from its use. The electric furnace method allowed production to increase to the point phosphorus could be used in weapons of war. In World WarI it was used in incendiaries, smoke screens and tracer bullets.

Today phosphorus production is larger than ever, used as a precursor for various chemicals, in particular the pesticide glyphosate sold under the brand name Roundup. Production takes place at large facilities and is transported heated to liquid form.

Occurrence

Due to its reactivity to air and many other oxygen containing substances, phosphorus is not found free in nature but it is widely distributed in many different minerals. Phosphate rock, which is partially made of apatite (an impure tri-calcium phosphate mineral), is an important commercial source of this element. Large deposits of apatite are located in China, Russia, Morocco, Florida, Idaho, Tennessee, Utah, and elsewhere. There are however concerns over how long these phosphorus deposits will last. The United States will deplete its deposits around 2035. China and Morocco have the largest known deposits today, but they too will eventually be depleted. During that depletion there could be a serious problem for the world's food production since phosphorus is such an essential ingredient in fertilizers.

The white allotrope can be produced using several different methods. In one process, tri-calcium phosphate, which is derived from phosphate rock, is heated in an electric or fuel-fired furnace in the presence of carbon and silica. Elemental phosphorus is then liberated as a vapor and can be collected under phosphoric acid.

Precautions

Organic compounds of phosphorus form a wide class of materials, some of which are extremely toxic. Fluorophosphate esters are among the most potent neurotoxins known. A wide range of organophosphorus compounds are used for their toxicity to certain organisms as pesticides (herbicides, insecticides, fungicides etc) and weaponized as nerve agents. Most inorganic phosphates are relatively nontoxic and essential nutrients.

The allotrope white phosphorus should be kept under water at all times as it presents a significant fire hazard due to its extreme reactivity to atmospheric oxygen, and it should only be manipulated with forceps since contact with skin can cause severe burns. Chronic white phosphorus poisoning of unprotected workers leads to necrosis of the jaw called "phossy-jaw". Ingestion of white phosphorus may cause a medical condition known as "Smoking Stool Syndrome".

When the white form is exposed to sunlight or when it is heated in its own vapor to 250 °C, it is transmuted to the red form, which does not phosphoresce in air. The red allotrope does not spontaneously ignite in air and is not as dangerous as the white form. Nevertheless, it should be handled with care because it does revert to white phosphorus in some temperature ranges and it also emits highly toxic fumes that consist of phosphorus oxides when it is heated.

Upon exposure to elemental phosphorus, in the past it was suggested to wash the affected area with 2% copper sulfate solution to form harmless compounds that can be washed away. The manual suggests instead "a bicarbonate solution to neutralize phosphoric acid, which will then allow removal of visible WP. Particles often can be located by their emission of smoke when air strikes them, or by their phosphorescence in the dark. In dark surroundings, fragments are seen as luminescent spots." Then, "Promptly debride the burn if the patient's condition will permit removal of bits of WP which might be absorbed later and possibly produce systemic poisoning. DO NOT apply oily-based ointments until it is certain that all WP has been removed. Following complete removal of the particles, treat the lesions as thermal burns." As white phosphorus readily mixes with oils, any oily substances or ointments are disrecommended until the area is thoroughly cleaned and all white phosphorus removed.

Isotopes

Radioactive isotopes of phosphorus include:

^{32}P; a beta-emitter (1.71 MeV) with a half-life of 14.3 days which is used routinely in life-science laboratories, primarily to produce radiolabeled DNA and RNA probes, *e.g.* for use in Northern blots or Southern blots. Because the high energy beta particles produced penetrate skin and corneas, and because any ^{32}P ingested, inhaled, or absorbed is readily incorporated into bone and nucleic acids, OSHA requires that a lab coat, disposable gloves, and safety glasses or goggles be worn when working with ^{32}P, and that working directly over an open container be avoided in order to protect the eyes. Monitoring personal, clothing, and surface contamination is also required. In addition, due to the high energy of the beta particles, shielding this radiation with the normally used dense materials (*e.g.* lead), gives rise to secondary emission of X-rays via a process known as Bremsstrahlung, meaning braking radiation. Therefore shielding must be accomplished with low density materials, *e.g.* Plexiglas, acrylic, Lucite, plastic, wood, or water.

^{33}P; a beta-emitter (0.25 MeV) with a half-life of 25.4 days. It is used in life-science laboratories in applications in which lower energy beta emissions are advantageous such as DNA sequencing.

Compounds

Ammonium phosphate ($(NH_4)_3PO_4$)
Phosphine (Phosphorus Trihydride PH_3)
Calcium phosphate ($Ca_3(PO_4)_2$)
Calcium dihydrogen phosphate ($Ca(H_2PO_4)_2$)
Iron(III) phosphate ($FePO_4$)

Iron(II) phosphate ($Fe_3(PO_4)_2$)
Hypophosphorous acid (H_3PO_2)

Nature of Phosphorus in Soils

Phosphorus (P) is an essential element classified as a macronutrient because of the relatively large amounts of P required by plants. Phosphorus is one of the three nutrients generally added to soils in fertilizers. One of the main roles of P in living organisms is in the transfer of energy. Organic compounds that contain P are used to transfer energy from one reaction to drive another reaction within cells.

Adequate P availability for plants stimulates early plant growth and hastens maturity. Although P is essential for plant growth, mismanagement of soil P can pose a threat to water quality. The concentration of P is usually sufficiently low in fresh water so that algae growth is limited. When lakes and rivers are polluted with P, excessive growth of algae often results. High levels of algae reduce water clarity and can lead to decreases in available dissolved oxygen as the algae decays, conditions that can be very detrimental to game fish populations.

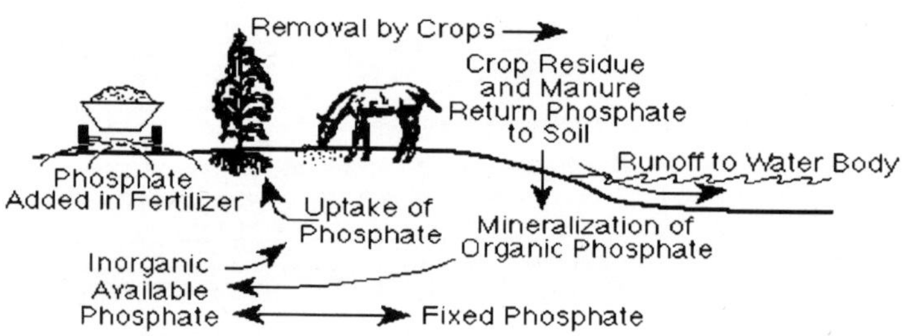

The phosphorus cycle.

Phosphate is taken up by plants from soils, utilized by animals that consume plants, and returned to soils as organic residues decay in soils. Much of the phosphate used by living organisms becomes incorporated into organic compounds. When plant materials are returned to the soil, this *organic phosphate* will slowly be released as inorganic phosphate or be incorporated into more stable organic materials and become part of the soil organic matter. The release of inorganic phosphate from organic phosphates is called *mineralization* and is caused by microorganisms breaking down organic compounds. The activity of microorganisms is highly influenced by soil temperature and soil moisture. The process is most rapid when soils are warm and moist but well drained. Phosphate can potentially be lost through soil erosion and to a lesser extent to water running over or through the soil.

Many phosphate compounds are not very soluble in water; therefore, most of the phosphate in natural systems exists in solid form. However, soil water and surface water (rivers and lakes) usually contain relatively low concentrations of *dissolved (or soluble) phosphorus.* Depending on the types of minerals in the area, bodies of water usually contain about 10 ppb or more of dissolved P as orthophosphate. Water bodies may also contain organic P and phosphate attached to small particles of sediment.

Total phosphorus in water is all of the phosphorus in solution regardless of its form and is often the form reported in water quality studies. *Algal available* or *bioavailable phosphorus* is P that is estimated to be available to organisms like algae that are present in a lake or river. This is usually estimated by a chemical test which is designed to measure the dissolved P and the particulate P that are easily available. This is a measure of the P that is of immediate concern to water quality.

Forms of Phosphorus in Soils

In soils P may exist in many different forms. In practical terms, however, P in soils can be thought of existing in 3 "pools":

Solution P
Active P
Fixed P

The *solution P pool* is very small and will usually contain only a fraction of a pound of P per acre. The solution P will usually be in the orthophosphate form, but small amounts of organic P may exist as well. Plants will only take up P in the orthophosphate form. The solution P pool is important because it is the pool from which plants take up P and is the only pool that has any measurable mobility. Most of the P taken up by a crop during a growing season will probably have moved only an inch or less through the soil to the roots. A growing crop would quickly deplete the P in the soluble P pool if the pool was not being continuously replenished.

The *active P pool* is P in the solid phase which is relatively easily released to the soil solution, the water surrounding soil particles. As plants take up phosphate, the concentration of phosphate in solution is decreased and some phosphate from the active P pool is released. Because the solution P pool is very small, the active P pool is the main source of available P for crops. The ability of the active P pool to replenish the soil solution P pool in a soil is what makes a soil fertile with respect to phosphate.

An acre of land may contain several pounds to a few hundred pounds of P in the active P pool. The active P pool will contain inorganic phosphate that is attached (or *adsorbed*) to small particles in the soil, phosphate that reacted with elements such as calcium or aluminum to form somewhat soluble solids, and organic P that is easily mineralized. Adsorbed phosphate ions are held on active sites on the surfaces of soil particles. The amount of phosphate adsorbed by soil increases as the amount of phosphate in solution increases and vice versa (*Figure 2*). Soil particles can act either as a source or a sink of phosphate to the surrounding water depending on conditions. Soil particles with low levels of adsorbed P that are eroded into a body of water with relatively high levels of dissolved phosphate may adsorb phosphate from the water, and vice versa.

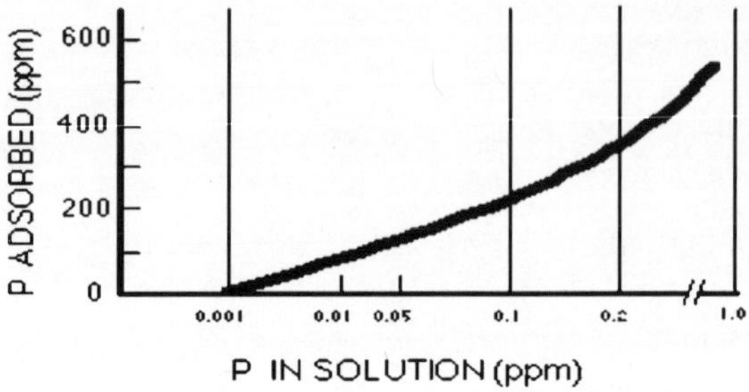

Fig . Relationship between P adsorbed by soil and P in solution.

The *fixed P pool* of phosphate will contain inorganic phosphate compounds that are very insoluble and organic compounds that are resistant to mineralization by microorganisms in the soil. Phosphate in this pool may remain in soils for years without being made available to plants and may have very little impact on the fertility of a soil. The inorganic phosphate compounds in this fixed P pool are more crystalline in their structure and less soluble than those compounds considered to be in the active P pool. Some slow conversion between the fixed P pool and the active P pool does occur in soils.

Fate of Phosphorus Added to Soils

The phosphate in fertilizers and manure is initially quite soluble and available. Most phosphate fertilizers have been manufactured by treating rock phosphate (the phosphate-bearing mineral that is mined) with acid to make it more soluble. Manure contains soluble phosphate, organic phosphate, and inorganic phosphate compounds that are quite available.

When the fertilizer or manure phosphate comes in contact with the soil, various reactions begin occurring that make the phosphate less soluble and less available. The rates and products of these reactions are dependent on such soil conditions as pH, moisture content, temperature, and the minerals already present in the soil.

As a particle of fertilizer comes in contact with the soil, moisture from the soil will begin dissolving the particle. Dissolving of the fertilizer increases the soluble phosphate in the soil solution around the particle and allows the dissolved phosphate to move a short distance away from the fertilizer particle. Movement is slow but may be increased by rainfall or irrigation water flowing through the soil. As phosphate ions in solution slowly migrate away from the fertilizer particle, most of the phosphate will react with the minerals within the soil. Phosphate ions generally react by adsorbing to soil particles or by combining with elements in the soil such as calcium (Ca), magnesium (Mg), aluminum (Al), and iron (Fe), and forming compounds that are solids. The adsorbed phosphate and the newly formed solids are relatively available to meet crop needs.

Gradually reactions occur in which the adsorbedphosphate and the easily dissolved compounds of phosphate form more insoluble compounds that cause the phosphate to be become fixed and unavailable. Over time this results in a decrease in soil test P. The mechanisms for the changes in phosphate are complex and involve a variety of compounds.

In alkaline soils (soil pH greater than 7) Ca is the dominant cation (positive ion) that will react with phosphate. A general sequence of reactions in alkaline soils is the formation of dibasic calcium phosphate dihydrate, octocalcium phosphate, and hydroxyapatite.

The formation of each product results in a decrease in solubility and availability of phosphate. In acidic soils (especially with soil pH less than 5.5) Al is the dominant ion that will react with phosphate. In these soils the first products formed would be amorphous Al and Fe phosphates, as well as some Ca phosphates. The amorphous Al and Fe phosphates gradually change into compounds that resemble crystalline variscite (an Al phosphate) and strengite (an Fe phosphate). Each of these reactions will result in very insoluble compounds of phosphate that are generally not available to plants. Reactions that reduce P availability occur in all ranges of soil pH but can be very pronounced in alkaline soils (pH > 7.3) and in acidic soils (pH < 5.5). Maintaining soil pH between 6 and 7 will generally result in the most efficient use of phosphate (*Figure*).

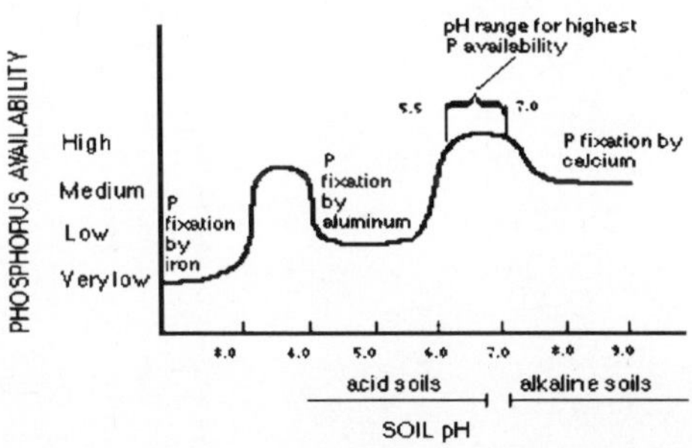

Fig. The availability of phosphorus is affected by soil pH.

Adding to the active P pool through fertilization will also increase the amount of fixed P. Depleting the active pool through crop uptake may cause some of the fixed P to slowly become active P.

The conversion of available P to fixed P is partially the reason for the low efficiency of P fertilizers. Most of the P fertilizer applied to the soil will not be utilized by the crop in the first season. Continued application of more P than the crops utilize increases the fertility of the soil, but much of the added P becomes fixed and unavailable.

Most fine-to medium-textured soils have large capacities to hold phosphate by adsorption and precipitation. Occasionally the question of how much phosphate a soil can hold is asked, especially when high loading rates of P are expected or have occurred. Soils differ in their phosphate holding capacity. Fine-textured soils can generally hold hundreds of pounds of phosphate per acre. while coarse-textured soils can generally hold much less phosphate due to the more inert character of sand particles as compared to clay particles. In addition the subsoil of many soils often has an even greater capacity to hold phosphate than does the corresponding surface soil. However, an important aspect of the ability of a soil to hold phosphate is that a soil cannot hold increasing amounts of phosphate in the solid phase without also increasing soil solution phosphate Figure . Increased amounts of phosphate in solution will potentially cause more phosphate to be lost to water running over the soil surface or leaching through the soil. Loading soils with very high levels of phosphate will generally not hurt crops but may result in increased phosphate movement to nearby bodies of water.

Predicting the Availability of Phosphorus in Soils

Soils may contain several hundred to several thousand pounds of phosphate per acre. However, much of the phosphate in soils is not available to growing plants. Phosphate in the soil solution P pool is immediately available but the amount is very small in comparison to the total P in soils. The active P pool is phosphorus that can be released into solution but is generally small in comparison to the fixed P.

To determine the need for supplemental P, soil tests are often used to estimate how much phosphate will be available for a crop. The most common way of determining P availability is to mix a small amount of soil with an extracting solution that contains an acid and/or complexing agent that will remove some of the phosphate from the soil particles. The extracting solution and soil are separated by filtration and the amount of P extracted is determined. In alkaline soils, a basic solution may be used as the extractant because an acidic solution will be neutralized by the alkaline soil and be less effective in extracting P. Calibration studies have been done to correlate crop response to fertilizer additions in soils with various soil test levels of P. Using the calibration data, recommendations can be made as to the amounts of phosphate fertilizer that will most likely give optimum yields.

Soil Phosphorus and Water Quality

Phosphorus is a somewhat unique pollutant in that it is an essential element, has low solubility, and is not toxic itself, but may have detrimental effects on water quality at quite low concentrations. There is considerable concern about P being lost from soils and transported to nearby streams and lakes. Several chemical properties of soil P have important implications for the potential loss of P to surface water.

Phosphorus in soils is almost entirely associated with soil particles. When soil particles are carried to a river or lake, P will be contained in this sediment. When the sediment reaches a body of water it may act as a sink or a source of P in solution. In either case, it is a potential source of P that may eventually be released.

Most soils have a large capacity to retain P. Even large additions of P will be mostly retained by soils provided there is adequate contact with the soil.

Increasing the amounts of phosphate in soils results in increased levels of phosphate in soil solutions. This will generally result in small but potentially important increases in the amounts of phosphate in water that passes over or through soils.

Phosphate in soils is associated more with fine particles than coarse particles. When soil erosion occurs, more fine particles are removed than coarse particles, causing sediment leaving a soil through erosion to be enriched in P.

Phosphorus Concerns in the Environment

Eutrophication can be caused by the nutrient enrichment of a water body. Nutrient movement in runoff and erosion from agricultural nonpoint sources is a resource management concern. The movement of phosphorus in runoff from agricultural land to surface water can accelerate eutrophication. Undesirable aquatic plant growth results from additions of phosphorus to the water. The net result of the eutrophic condition and excess plant growth in water is the depletion of oxygen in the water due to the heavy oxygen demand by microorganisms as they decompose the organic material. Little attention has been given to management strategies to minimize the nonpoint movement of P in the landscape because of the easier identification and control of point source inputs of P to surface waters and a lack of direct human health risks associated with eutrophication. Phosphorus is generally the limiting nutrient in fresh water systems and any increase in P is usually results in more aquatic vegetation. Society is concerned about maintaining clean drinking water. This concern now includes a cost for removing the color, taste, and odor associated with the high trophic condition and vegetation growth in surface water due to excess nutrients.

Phosphorus Movement in the Landscape

Phosphorus movement in runoff occurs as particulate P and dissolved P. Particulate P is attached to mineral and organic sediment as it moves with the runoff. Dissolved P is in the water solution. In general, particulate P is the major portion (75-90%) of the P transported in runoff from cultivated land. Dissolved P makes up a larger portion of the total P in runoff from non- cultivated lands such as pastures and fields with reduced tillage. In terms of their impact on eutrophication of water bodies, particulate P becomes less available to algae and plant uptake than dissolved P because of the chemical form it has with the mineral (particularly iron, aluminum, and calcium) and organic compounds. The availability of particulate P to plants and algae is variable, ranging from 10 to 90% of the total P. yet can represent a long-term source of P for algae and plant uptake from the water body. Dissolved P is 100% bioavailable to plants. Added together, the bioavailable portion of particulate P and the dissolved P represents the phosphorus that promotes eutrophication of surface waters.

The method by which P in both particulate and dissolved form moves within the landscape are simplified in the following description. Eroding soil material is transported by runoff. During detachment and movement of sediment in runoff, the finer clay-sized fraction of the source material are preferentially eroded. The P content and reactivity of the eroded material to P are usually greater than the source soil from which it was eroded. The suspended sediment in the runoff can rapidly adsorb the dissolved P in the runoff water.

As runoff moves from the landscape and toward the water body there is generally a progressive dilution of P through additions of water and a reduction in the amount of sediment carried because of sediment deposition. Phosphorus may become more bioavailable by the sorption and Resorption processes and by the preferential transport of claysized material as sediment moves over the landscape.

The movement of dissolved P begins with the Resorption, dissolution, and extraction of P from the soil, plant, and organic material. These processes occur when rain and runoff water interact with the thin layer of surface soil (0.05 to 0.10 inches). Some water infiltrates into the soil and percolates through the profile where desorption of P will result in a low dissolved concentration in subsurface and return flow. High dissolved P concentration can be expected in the water percolating through organic, coarse-textured, and oxygen depleted (reduced), water-logged soils. Soil pH also affects the movement and availability of phosphorus.

The interaction between the particulate and dissolved P in the runoff is very dynamic and the mechanism of transport is complex. Therefore, it is difficult to predict the transformation and ultimate fate of P as it moves through the landscape.

The Concept

The purpose of the Phosphorus Index is to provide field staffs, watershed planners, and land users with a tool to assess the various landforms and management practices for potential risk of phosphorus movement to water bodies. The ranking of Phosphorus Index identifies sites where the risk of phosphorus movement may be relatively higher than that of other sites. When the parameters of the index are analyzed, it will become apparent that an individual parameter or parameters may be influencing the index disproportionately. For the first version, the Phosphorus Index will be a 8 x 5 matrix using a limited number of landform site and management characteristics. The input to the matrix will be readily accessible field data. This index will be used as a tool for understanding the contribution that individual landform and management parameters have toward risk of phosphorus movement and will provide a method for developing management guidelines for phosphorus at the site to lessen their impact on water quality.

The P Index is a simple 8 by 5 matrix that relates site characteristics with a range of value categories. This first version (Table) of the P Index has eight site characteristics. The eight characteristics are:

- Soil erosion
- Irrigation erosion
- Runoff class
- Soil P test
- P fertilizer application rate
- P fertilizer application method
- Organic P source application rate
- Organic P source application method
- The five value categories are:
- None
- Low
- Medium
- High
- Very high

There are eight site characteristics used in the P Index to assess a particular site. Each site characteristic is rated NONE, LOW, MEDIUM, HIGH, or VERY HIGH by determining the range for each category. Although ranges have been given in this version of the index, it is imperative that each user establish their own range of values for each characteristic. Each user group should consider the cropping, soil, management conditions in the area designated to use the Phosphorus Index. Consideration need to be made as to base conditions for soil erosion, soil phosphorus test levels, and crop requirements. From these local considerations the ranges of the values category can be made. Any user group should include NRCS, Extension Service, University and ARS researchers, as well as state natural resource agencies and other groups interested in water quality.

The definition of each of the eight site characteristics are:

Soil Erosion

Soil erosion is defined as the loss of soil along the slope or unsheltered distance caused by the processes of water and wind. Soil erosion is estimated from erosion prediction models currently used (USLE or RUSLE for water erosion and WEQ for wind erosion). Erosion induced by irrigation is calculated by other convenient methods. The value category is given in tons of soil loss per acre per year (ton/acre/year). These soil loss prediction models do not predict sediment transport and delivery to a water body. The prediction models are used in this index to indicate a movement of soil, thus potential for sediment and attached phosphorus movement across the slope or unsheltered distance and toward a water body.

Irrigation Erosion

Potential P loss resulting from furrow irrigation induced erosion is considered by inclusion of a rating system based on soil susceptibility to particle detachment by hydraulic shear and flow rate of water in the furrow. The susceptibility to detachment is given by a relative ranking of soil erodibility classes under furrow irrigation (table). These classes are an initial attempt at a relative ranking based on inherent stable and static soil properties (i.e., texture and clay mineralogy). There are temporal variations in the relative erodibility and actual amount of erosion with furrow erosion. These changes in erodibility are a function in the soil properties or a result of management. However, no attempt is made to consider temporal soil properties nor management factors in the rating. The introduced flow rate in the furrow (Q) is given by the irrigation plan and recorded as gallons per minute (gal/min). The furrow slope (S) of the site is given as a percentage (feet per 100 feet). The product of flow rate (Q) and slope (S) is used to determine the value category.

Runoff Class

The runoff class of the site can be determined from soil survey data. Guidance in determining the runoff class is based on the soil saturated hydraulic conductivity (K_{Sat}) and the percent slope of the site. Another method to determine runoff class is using the NRCS curve number (CN) method. The major factors that determine curve number are the hydrologic soil group. Landscape cover type, conservation treatment, hydrologic condition, and antecedent runoff conditions.

Soil P Test

A soil sample from the site is necessary to assess the level of "available P" in the surface layer of the soil. The available P is the level customarily given in a sail test analysis by the Cooperative Extension Service or commercial soil test laboratories. The user of the P Index must determine the ranges of soil test P in each value category (LOW, MEDIUM, HIGH, and EXCESSIVE). These ranges of soil test P values will vary by soil test method and region. The soil test level for "available P" does not ascertain the total P in the surface soil. It does however, give an indication of the amount of total P that may be present because of the general relationship between the forms of P (organic, adsorbed, and labile P) and the solution P available for crop uptake.

P Fertilizer Application Rate

The P fertilizer application rate is the amount, in pounds per acre (lb/acre), of phosphate fertilizer (P_20_5) that is applied to the soil. This phosphate fertilizer does not include phosphorus from organic sources.

P Fertilizer Application Method

The manner in which P fertilizer is applied to the soil and the amount of time that the fertilizer is exposed on the soil surface until crop utilization effects potential P movement. Incorporation implies that the fertilizer P is buried below the soil surface at a minimum of two inches. The value categories of increasing severity, LOW to VERY HIGH, depict the longer surface exposure time between fertilizer application, incorporation, and crop utilization.

Organic P Source Application Rate

The organic P application rate is the amount, in pounds per acre (lb/acre), of potential phosphate (P_2O_5) that is contained in the manure and applied to the soil. An analysis of the organic material is necessary to determine the potential phosphate content of the manure.

The P content by analysis is generally considered to be completely plant available. This organic phosphate source does not include phosphorus from fertilizer sources .

Organic P Source Application Method

The manner in which organic P material is applied to the soil and the time that the organic material is exposed on the soil surface until crop utilization can determine potential P movement. Incorporation implies that the organic P material is buried below the soil surface at a minimum of two inches. The value categories of increasing severity, LOW to VERY HIGH, depict the longer surface exposure time between organic P material application, incorporation, and crop utilization.

The Procedures for Making an Assessment

The site characteristics have been assigned a weighting based on the reasoning that particular site characteristics may be more prominent than others in allowing potential phosphorus movement from the site. There is scientific basis for concluding that these relative differences exist; however, the absolute weighting factors given are based currently on professional judgment. The site characteristic weighting factors are:

- Soil erosion (1.5)
- Irrigation erosion (1.5)
- Runoff class (0.5)
- Soil P test (1.0)
- P fertilizer application rate (0.75)
- P fertilizer application method (0.5)
- Organic P source application rate (1.0)
- Organic P source application method (1.0)

The value categories are rated using a log base of 2. The greater ratings, the proportionally higher are the values. The higher the value, the higher potential for significant problems related to phosphorus movement.

The value ratings are:

- none = 0
- low = 1
- medium = 2
- high = 4
- very high = 8

To make an assessment using the P Index, select a rating value for each site characteristic using the categories NONE, LOW, MEDIUM, HIGH, or VERY HIGH. Multiply the site characteristic weight factor by the rating value to get the weighted value for that characteristic. Proceed to rate and factor each characteristic of the index. Sum the weighted values for all eight characteristics, and compare the total with site vulnerability chart.

A description of vulnerability is given to appraise the assessment of the P Index.

Interpretations of Vulnerability Rating. for the P Index

Low - This site has a Low potential for P movement from the site. If farming practices are maintained at current level, the probability of an adverse impact to surface water resources from P losses from this site would be low.

Medium - This site has a Medium potential for P movement from the site. The probability for an adverse impact to surface water resources is greater than that from a LOW vulnerability rated site. Some remedial action should be taken to lessen the probability of P movement.

High - This site has a High potential for P movement from the site. There is a high probability for an adverse impact to surface water resources unless remedial action is taken. Soil and water conservation as well as phosphorus management practices are necessary to reduce the risk of P movement and probable water quality degradation.

Very High - This site has a Very High potential for P movement from the site. The probability for an adverse impact to surface water resources is very high. Remedial action is required to reduce the risk of P movement. All necessary soil and water conservation practices plus a phosphorus management plan must be put in place to reduce the potential of water quality degradation.

An Example in using the Phosphorus Index in Table:

Table. Phosphorus Index for assessing the vulnerability of a land unit. Summation of the weighted rating value is used to determine the site vulnerability.

soil erosion (weight = 1.5) is 7.5 ton/ac/yr (=MEDIUM, value = 2)	1.5 x 2 = 3.0
irrigation erosion (weight = 1.5) is not applicable (=NONE, value = 0)	1.5 x 0 = 0
runoff class (weight = 0.5) is LOW (value = 1)	0.5 x 1 = 0.5
soil P test (weight = 1.0) is 82 lb P (=HIGH, value = 4)	1.0 x 4 = 4.0
P fertilizer application rate (weight = 0.75) is 25 lb/ac (=LOW, value = 1)	0.75 x 1 = 0.75
P fertilizer application method (weight = 0.5) is placed with planter (=LOW, value = 1)	0.5 x 1 = 0.5
organic P source application rate (weight = 1.0) is 95 lb/ac (=VERY HIGH, value = 8)	1.0 x 8 = 8.0
organic P source application method (weight = 1.0) is surface applied a month before notill planting (=HIGH, value =4)	1.0 x 4 = 4.0

Sum Total of all weighted values = 20.75

Site Vulnerability is HIGH

HIGH - This site has a HIGH potential for P movement from the site. There is a high probability for an adverse impact to surface water resources unless remedial action is taken. Soil and water conservation as well as phosphorus management practices are necessary to reduce the risk of P movement and probable water quality degradation.

I. QS value

Q = flow rate of water introduced into the furrow (in gallons per minute, GPM).

S = furrow slope (in feet per 100 feet, percent).

Example: For a 5 gpm flow rate and a 2% furrow grade:

QS = 5 gpm * 2% grade = 10

II. Relative ranking of soil erodibility under furrow irrigation

Use local criteria to determine the relative erodibility of the soil in question. If no local criteria are established, use the following for guidance:

A. Very Erodible Soils

Soils in which the surface layer texture is silt, or silt loam with < 15% nonmontmorillonitic clay, or fine and very fine sandy loam with < 15% nonmontmorillonitic clay, or loamy fine sand, or loamy very fine sand. Contact a soil scientist for clay content and mineralogy.

B. Erosion-resistant Soils

Soils that have the following characteristics in the upper 5 cm of the surface layer:

silty clay, clay, or sandy clay texture, weak or massive structure, and mixed or montmorillonitic clay mineralogy.

other soils that have medium or coarse blocky structure or coarse granular structure (i.e. natural aggregates > 10 mm) and very firm or firmer rupture resistance class in the moist state (i.e. requires at least strong force between thumb and forefinger to cause failure of a moist soil aggregate).

C. Erodible Soils

Soils that have a surface layer not fitting any of the above criteria.

Table. Furrow IRRIGATION EROSION site characteristics

Site characteristic (weight)	Phosphorus loss rating (value)				
	None (0)	Low (1)	Medium (2)	High (4)	Very high (8)
Soil erosion (1.5)	Not applicable	<5 tons/ac	5-10 tons/ac	10-15 tons/ac	>15 tons/ac
Irrigation erosion (1.5)	Not applicable	Tailwater recovery or qs<6 for very erodible soils or qs<10 for other soils	Qs>10 for erosion resistant soils	Qs>10 for erodible soils	Qs>6 for very erodible soils
Runoff class (0.5)	Negligible	Very low or low	Medium	High	Very high
Soil p-test (1.0)	Not applicable	Low	Medium	High	Excessive
P-fertilizer application rate (0.75)	None applied	1-30 p_2o_5 lbs/ac	31-90 p_2o_5 lbs/ac	91-150 p_2o_5 lbs/ac	> 150 p_2o_5 lbs/ac
P-fertilizer application method (0.5)	None applied	Placed with planter deeper than 2 inches	Incorporated immediately before crop	Incorporated > 3 months before crop or surface applied < 3 months before crop	Surface applied > 3 months before crop
Organic p source application rate (1.0)	None applied	1-30 p_2o_5 lbs/ac	31-60 p_2o_5 lbs/ac	61-90 p_2o_5 lbs/ac	> 90 p_2o_5 lbs/ac
Organic p source application method (1.0)	None	Injected deeper than 2 inches	Incorporated immediately before crop	Incorporated > 3 months before crop or surface applied < 3 months before crop	Surface applied to pasture, or > 3 months before crop

Total of Weighted Rating Values	Site Vulnerability
< 8	LOW
8 -14	MEDIUM
15 - 32	HIGH
> 32	VERY HIGH

Sulphur

Sulphur or sulphur is the chemical element in the periodic table that has the symbol S and atomic number 16. It is an abundant, tasteless, odorless, multivalent non-metal. Sulphur, in its native form, is a yellow crystaline solid. In nature, it can be found as the pure element or as sulfide and sulfate minerals. It is an essential element for life and is found in two amino acids. Its commercial uses are primarily in fertilizers but it is also widely used in gunpowder, matches, insecticides and fungicides.

Notable characteristics

A piece of sulphur melts to a blood-red liquid. When burned, it emits a blue flame.

At room temperature, sulphur is a soft bright yellow solid. Although sulphur is infamous for its smell—frequently compared to rotten eggs—the odor is actually characteristic of hydrogen sulfide (H_2S); elemental sulphur is odorless. It burns with a blue flame that emits sulphur dioxide, notable for its peculiar suffocating odor. Sulphur is insoluble in water but soluble in carbon disulfide and to a lesser extent in other organic solvents such as benzene. Common oxidation states of sulphur include "2, +2, +4 and +6. Sulphur forms stable compounds with all elements except the noble gases.

Applications

Sulphuric acid production is the major end use for sulphur, and consumption of sulphuric acid has been regarded as one of the best indices of a nation's industrial development. Sulphur is also used in batteries, detergents, the vulcanization of rubber, fungicides, and in the manufacture of phosphate fertilizers. Sulfites are used to bleach paper and as a preservative in wine and dried fruit.

Because of its flammable nature, sulphur also finds use in matches, gunpowder, and fireworks. Sodium or ammonium thiosulfate are used as photographic fixing agents. Magnesium sulfate, better known as Epsom salts, can be used as a laxative, a bath additive, an exfoliant, or a magnesium supplement for plants. Sulphur is used as the light-generating medium in the rare lighting fixtures known as sulphur lamps.

Biological role

The amino acids cysteine and methionine contain sulphur, as do all polypeptides, proteins, and enzymes which contain these amino acids. This makes sulphur a necessary component of all living cells. Disulfide bonds between polypeptides are very important in protein assembly and structure. Homocysteine and taurine are also sulphur containing amino acids but are not coded for by DNA nor are they part of the primary structure of proteins. Some forms of bacteria use hydrogen sulfide (H_2S) in the place of water as the electron donor in a primitive photosynthesis-like process. Sulphur is absorbed by plants via the roots from soil as the sulfate ion and reduced to sulfide before it is incorporated into cysteine and other organic sulphur compounds (sulphur assimilation). Inorganic sulphur forms a part of iron-sulphur clusters, and sulphur is the bridging ligand in the Cu_A site of cytochrome c oxidase. Sulphur is an important component of coenzyme A.

Environmental impact

The burning of coal and petroleum by industry and power plants liberates huge amounts of sulphur dioxide (SO_2) which reacts with atmospheric water and oxygen to produce sulphuric acid. This sulphuric acid is a component of acid rain, which lowers the pH of soil and freshwater bodies, resulting in substantial damage to the natural environment and chemical weathering of statues and architecture. Fuel standards increasingly require sulphur to be extracted from fossil fuels to prevent the formation of acid rain. This extracted sulphur is then refined and represents a large portion of sulphur production.

History

Homer mentioned "pest-averting sulphur" in the 8th century BC and in 424 BC, the tribe of Boeotia destroyed the walls of a city by burning a mixture of coal, sulphur, and tar under them. Sometime in the 12th century, the Chinese invented gun powder which is a mixture of potassium nitrate (KNO_3), carbon, and sulphur. Early alchemists gave sulphur its own alchemical symbol which was a triangle at the top of a cross. In the late 1770s, Antoine Lavoisier helped convince the scientific community that sulphur was an element and not a compound. In 1867, sulphur was discovered in underground deposits in Louisiana and Texas. The overlying layer of earth was quicksand, prohibiting ordinary mining operations. Therefore the Frasch process was utilized.

Occurrence

Elemental sulphur can be found near hot springs and volcanic regions in many parts of the world, especially along the Pacific Ring of Fire. These occurrences are the basis for the traditional name brimstone, since sulphur could be found near the brims of volcanic craters. Such volcanic deposits are currently exploited in Indonesia, Chile, and Japan.

Significant desposits of elemental sulphur also exist in salt domes along the coast of the Gulf of Mexico, and in evaporites in eastern Europe and western Asia. The sulphur in these deposits is believed to come from the action of anaerobic bacteria on sulfate minerals, especially gypsum. Such deposits are the basis for commercial production in the United States, Poland, Russia, Turkmenistan, and Ukraine.

Sulphur extracted from oil, gas and the Athabasca Oil Sands has become a glut on the market, with huge stockpiles of sulphur in existence throughout Alberta.

Compounds

Many of the unpleasant odors of organic matter are based on sulphur-containing compounds such as methyl and ethyl mercaptan used to scent natural gas so that leaks are easily detectable. The odor of garlic and "skunk stink" are also caused by sulphur-containing organic compounds. However, not all organic sulphur compounds smell unpleasant; for example, grapefruit mercaptan, a sulphur-containing monoterpenoid is responsible for the characteristic scent of grapefruit.

Polymeric sulphur nitride has metallic properties even though it does not contain any metal atoms. This compound also has unusual electrical and optical properties. This polymer can be made from tetrasulphur tetranitride S_4N_4.

Inorganic sulphur compounds:

- Sulfides (S^{2-}), a complex family of compounds usually derived from S^{2-}. Cadmium sulfide (CdS) is an example.
- Sulfates (SO_4^{2-}), the salts of sulphuric acid. Sulphuric acid also reacts with SO_3 in equimolar ratios to form pyrosulphuric acid ($H_2S_2O_7$).
- Sodium dithionite, $Na_2S_2O_4$, is the highly reducing dianion derived from hyposulphurous/dithionous acid.
- Sodium dithionate ($Na_2S_2O_6$).
- Polythionic acids ($H_2S_nO_6$), where n can range from 3 to 80.
- Sodium polysulfides (Na_2S_x)
- Sulphur hexafluoride, SF_6, a dense gas at ambient conditions, is used as nonreactive and nontoxic propellant
- Sulphur nitrides are chain and cyclic compounds containing only S and N. Tetrasulphur tetranitride S_4N_4 is an example.

Isotopes

Sulphur has 18 isotopes, of which four are stable: ^{32}S (95.02%), ^{33}S (0.75%), ^{34}S (4.21%), and ^{36}S (0.02%). Other than ^{35}S, the radioactive isotopes of sulphur are all short lived. ^{35}S is formed from cosmic ray spallation of ^{40}Ar in the atmosphere. It has a half-life of 87 days.

When sulfide minerals are precipitated, isotopic equilibration among solids and liquid may cause small differences in the äS-34 values of co-genetic minerals. The differences between minerals can be used to estimate the temperature of equilibration. The äC-13 and äS-34 of coexisting carbonates and sulfides can be used to determine the pH and oxygen fugacity of the ore-bearing fluid during ore formation.

In most forest ecosystems, sulfate is derived mostly from the atmosphere; weathering of ore minerals and evaporites also contribute some sulphur. Sulphur with a distinctive isotopic composition has been used to identify pollution sources, and enriched sulphur has been added as a tracer in hydrologic studies.

Differences in the natural abundances can also be used in systems where there is sufficient variation in the ^{34}S of ecosystem components. Rocky Mountain lakes thought to be dominated by atmospheric sources of sulfate have been found to have different äS-34 values from lakes believed to be dominated by watershed sources of sulfate.

Precautions

Carbon disulfide, carbon oxysulfide, hydrogen sulfide, and sulphur dioxide should all be handled with care.

Although sulphur dioxide is sufficiently safe to be used as a food additive in small amounts, at high concentrations it reacts with moisture to form sulphurous acid which in sufficient quantities may harm the lungs, eyes or other tissues. In creatures without lungs such as insects or plants, it otherwise prevents respiration.

Hydrogen sulfide is quite toxic (more toxic than cyanide). Although very pungent at first, it quickly deadens the sense of smell, so potential victims may be unaware of its presence until it is too late.

Effect of Sulphur on Wheat Plant Growth

Sulphur is a building block of protein and a key ingredient in the formation of chlorophyll. Without adequate sulphur, crops can not possibly reach their full potential in terms of yield or protein content. Nor can they make efficient use of nitrogen, phosphorus and other vital elements.

Sulphur Content and Requirement of Wheat Plants

Wheat plants have a lesser requirement for sulphur than oilseed crops such as canola. A 2690 kg/ha (40 bu/ac) wheat crop will contain in the seed and straw 12 kg/ha (10 lb/ac) of sulphur, of which only 40% is in the seed.

		Sulphur (S)	
Yield	Crop Part	Kg/ha	(lb/ac)
Wheat	Seed	5	4
2690 Kg/ha	Straw	7	6
(40 Bu/ac)	Total	12	10
Canola	Seed	13	12
1960 Kg/ha	Straw	10	9

Time of application

If sulphate fertilizers are applied in the fall, a portion of the sulphate could leach downward as much as 50 cm (20 in), depending on the amount of fall and spring precipitation and the texture of the soil. Therefore, sulphate fertilizer is usually most efficient when applied in the spring.

Elemental sulphur fertilizer applied in the spring of the crop year may not oxidize to sulphate sufficiently or early enough to supply the crop needs.

Farmers who anticipate sulphur deficiencies and plan to utilize elemental S, should apply 10-20 lb/ac of elemental sulphur on an annual to biannual basis to avoid deficiencies.

Potassium, calcium and magnesium

Potassium

Potassium is a chemical element in the periodic table. It has the symbol K (L. *kalium*) and atomic number 19. The name "potassium" comes from the word "potash", as potassium was first isolated from potash. Potassium is a soft silvery-white metallic alkali metal that occurs naturally bound to other elements in seawater and many minerals. It oxidizes rapidly in air, is very reactive, especially in water, and resembles sodium chemically.

Calcium

Calcium is the chemical element in the periodic table that has the symbol Ca and atomic number 20. Calcium is a soft grey alkaline earth metal that is used as a reducing agent in the extraction of thorium, zirconium and uranium. Calcium is also the fifth most abundant element in the Earth's crust. It is essential for living organisms, particularly in cell physiology, and is the most common metal in many animals.

The most abundant isotope, ^{40}Ca, has a nucleus of 20 protons and 20 neutrons. Its electron configuration is: 2 electrons in the K shell (principal quantum number 1), 8 in the L shell (principal quantum number 2), 8 in the M shell (principal quantum number 3), and 2 in the N shell (principal quantum number 4). The outer shell is the valence shell, with 2 electrons in the lone 4s orbital, the 3p orbitals being empty.

Calcium is not naturally found in its elemental state. Calcium is found mostly in soil systems as limestone, gypsum and fluorite. Stalagmites and stalactites contain calcium carbonate. Being an essential macromineral in the human diet, soil conservation practices often consider the sustainable equilibrium of calcium concentrations in the earth.

All soils contain calcium ions (Ca_2+) and magnesium (Mg_2+) cations (positively charged ions) attracted to the negative exchange sites on clays and organic matter (cation exchange complex of the soil). The amount and relative proportion usually reflect the soil's parent materials. Calcium (Ca) and Mg are plant-essential nutrients, and the ionic form of each held on the soil exchange sites is the form taken up by plants.

The usual approach for determining whether the soil supply is sufficient to meet crop needs is to extract soil with 1 molar (M) ammonium acetate (the same procedure used to determine soil test potassium) and evaluate the amount measured against critical levels. Because Iowa soils contain more than adequate levels of these nutrients, no critical level has been or can be established. Therefore, exchangeable Ca and Mg are not routinely tested, nor are there Iowa State University publications that give soil test Ca or Mg interpretations. Unless you are interested in the soil cation exchange capacity (CEC)—in routine soil testing determined by summing the dominant exchangeable cations (Ca_2+, Mg_2+, K+, H+)—soil samples don't need to be analyzed in the laboratory for Ca and Mg from most Iowa fields. Also, Iowa soils have large quantities of both nutrients and they are replenished by limestone application.

How is a Ca:Mg ratio calculated?

Once exchangeable Ca and Mg are determined by laboratory analysis, the ratio is calculated using the meq basis (electrical charge basis). For example, if there is 4.88 meq Ca/100 g soil and 1.72 meq Mg/100 g, then the Ca:Mg ratio is 2.8:1. Table gives the exchangeable Ca, Mg, and calculated Ca:Mg ratio for several Iowa soils. These values are typical for Iowa soils. Soil Ca:Mg ratios naturally are above 1:1.

Why the interest in Ca:Mg ratios?

From the above-mentioned statement that Ca and Mg levels are higher than needed for crop production in Iowa soils, you can easily conclude that ignoring the ratio is just fine. Research confirms that this conclusion is justified; however, promotion of the ratio concept persists today despite many years of research that indicates otherwise.

The origin of this concept was derived from work by Bear and colleagues in the 1940s. However, their work did not differentiate between crop response (alfalfa) due to pH improvement from lime application to acid soils and the change in Ca:Mg. Other research during the same time indicated that ratios were not important. Many research trials since have not shown an influence of Ca:Mg ratio on crop production. Conclusions from the researchers were "The results strongly suggest that for maximum crop yields, emphasis should be placed on providing sufficient, but nonexcesive levels of each basic cation rather than attempting to attain a favorable basic cation saturation ratio (BCSR), which evidently does not exist."

Various greenhouse and field trials indicate that crop productivity is not influenced by ranges from less than 1:1 to more than 25:1—ratios outside of what is normally measured in soils. Plants also play a role in Ca and Mg uptake and exclude excess Ca or Mg at the root surface.

In summary, the Ca:Mg ratio concept is unproven and should not be used as a basis for fertilization or liming practices. Having sufficient levels of Ca and Mg is the proper method of evaluation, rather than trying to manipulate ratios. We are fortunate in Iowa that soil Ca and Mg levels are normally adequate, and maintenance of plant-available Ca and Mg occurs either because the soil has a large supply or because of liming with local quarry limestone to maintain adequate soil pH for crop production.

Magnesium

Magnesium is an essential ingredient of chlorophyll, the green plant pigment that gives leaves their colour and enables the plant to make food from sunlight.

Magnesium is essential for enzymes in plants and especially in chlorophyll – the green pigment that enables plant to carry out photosynthesis. Magnesium is a metallic element represented by the symbol Mg.

Soil pH

How well plants grow can be strongly influenced by whether the soil is acid or alkaline. It's best to get the pH right before dealing with other nutrient problems.

Lack of magnesium

This appears as a lack of green in the leaves. The area between the veins of leaves, especially the lower older ones, goes yellow.

Sometimes the leaf margins also lose their green colour. Plants of the cabbage family sometimes develop a purplish colour. Sandy soils are most likely to lack magnesium. This deficiency is also seen when there is plenty of magnesium in the soil, but the roots cannot take it up because of poor soil structure, drought or water-logging. When the underlying problems are resolved the magnesium deficiency isappears.

Fast-growing plants may show a temporary magnesium deficiency in the older leaves. This usually sorts itself out when the plant begins to grow more slowly. There is no need to add extra magnesium. Magnesian limestone is acceptable to organic growers.

Common magnesium fertilisers

The amount of magnesium in fertiliser has to be stated on the packet. However, few compound fertilisers contain it, as plants need much less magnesium than nitrogen, phosphorus and potassium. Both Epsom salts (10% Mg) and kieserite (16% Mg) are fast-acting, water-soluble fertilisers containing magnesium sulphate.

Kieserite contains less water in its crystals than the morefrequently sold Epsom salts and is usually cheaper to use when treating large areas. Because it is less soluble, it can be used in seedbeds, whereEpsom salts could scorch delicate young roots. Epsom salts are used for foliar feeding, as they are more soluble in water than ieserite. Neither fertiliser is organic. Magnesium is easily washed out of soils that contain little or no clay. Since these soils are often acid and need lime, maintain the pH using magnesian limestone. This ground limestone, derived from magnesium-rich rocks, is also known as Maglime or Dolodust. but do not use it where you intend to grow acid-loving plants like rhododendrons and heathers. Foliar-feed these plants instead –especially rhododendrons which can suffer from magnesium deficiency.

Some liquid fertilisers contain a little magnesium – these are a useful pick-me-up for containergrown plants – tomatoes in growing bags, for example. Manure, seaweed, compost and mushroom compost contain small amounts of magnesium. Cattle manure, for example, contains 0.04%Mg.

Magnesium and Potassium

High levels of potassium (K) in the soil suppress plants' uptake of magnesium, so avoid overdosing with fertilisers containing potassium. Conversely, overfeeding with magnesium fertiliser can induce a potassium deficiency in plants. Foliar feeding helps overcome this imbalance.

General guidelines

If your soil has not been analysedrecently and you believe that plants are showing magnesium deficiency, try a foliar feed in the first instance. If this has some effect, the soil may be deficient, and applying a magnesium fertiliser should be worthwhile. Spreading 30g a sq m (1oz a sq).

Suppliers

Magnesian limestone, Epsom salts and kieserite:

Bulky organic manures

Where the magnesium level is low or very low, manure, mushroom compost or garden compost are unlikely to provide enough and you will need fertilisers to boost levels for good growth.

Where levels are high or very high, use good-quality manure or compost at 4.5kg a sq m (10lb a sq yd). Manures and composts are highly variable in composition, depending on what they have been made from, how they have been stored and how well-rotted they are. Also, they can vary greatly in water content. If in doubt, consider the organic matter as a soil conditioner and add fertiliser to provide nutrients.

Foliar feeding

Wetting the leaves of plants suffering from a magnesium deficiency with a solution of Epsom salts can provide adequate magnesium. However, you may need to foliar feed again during the growing season and every time vulnerable plants are grown. Susceptible perennials, for example, typically need three or four applications per year.

Spraying young foliage during warm, humid weather gives the best results. Foliar feeding doesn't work so well on droughtstressed plants with sparsefoliage.

Avoid foliar feeding during hot, dry weather as the solution may dry too quickly, and the resulting concentrate could scorch the leaves. If you're in doubt about the suitability of weather conditions, apply the feed in the evening.

Use 20g of Epsom salts per litre of water (3oz per gallon of water). Add a little washing-up liquid – this will act as a 'wetter' to stop the solution running straight off. Check if there is enough wetter by dipping a leaf in the solution – if it runs off without wetting the leaf, more is needed. Too much wetter can damage plants – if you're doing the whole of a precious plant, check by spraying a small shoot and leaving it for a few days before spraying the remainder.

Soil analysis

To find out how much magnesium your soil contains requires soil analysis. This should give you a 'magnesium level' from very low to very high, or a 'magnesium index' from 0 to 3. You can then use the tables to determine what to add. Alternatively, a Gardening Which? Soil Analysis will tell you the levels and recommend what action to take.

Trace Element

Trace element is a chemical element whose concentration is less than 1000 ppm or 0.1% of rocks composition.

Microminerals

Microminerals(also known as trace elements) are micronutrients that are chemical elements. They include boron, chromium, cobalt, copper, fluorine, iodine, iron, manganese, molybdenum, selenium, silicon, tin, vanadium, and zinc. They are dietary minerals needed by the human body in very small quantities (generally less than 100mg per day) as opposed to macrominerals which are required in larger quantities. Note that the use of the term "mineral" in this context is to be distinguished from the definition commonly used in the geological sciences.

Micronutrients for plants:

There are about eight nutrients essential to plant growth and health that are only present in very small quantities. These are manganese, boron, copper, iron, chlorine, cobalt, molybdenum, and zinc. Some consider sulphur a micronutrient, but it is listed here as a macronutrient. Though these are present in only small quantities, they are all necessary.

Boron - One of eight micronutrients. Boron is believed to be involved in carbohydrate transport in plants; it also assists in metabolic regulation. Boron deficiency will often result in bud dieback.

Chlorine - One of eight micronutrients in plants. Chlorine is necessary for osmosis and ionic balance; it also plays a role in photosynthesis.

Cobalt - One of eight micronutrients essential to plant health. Cobalt is thought to be an important catalyst in nitrogen fixation. It may need to be added to some soils before seeding legumes.

Copper - One of eight micronutrients in plants. It is a component of some enzymes and of vitamin A. Symptoms of copper deficiency include browning of leaf tips and chlorosis.

Iron - One of eight micronutrients in plants. It is essential for chlorophyll synthesis, which is why an iron deficiency results in chlorosis.

Manganese - One of eight micronutrients. Manganese activates some important enzymes involved in chlorophyll formation. Manganese deficient plants will develop chlorosis between the veins of its leaves. The availability of manganese is partially dependent on soil pH.

Molybdenum - One of eight micronutrients essential to plant health. Molybdenum is used by plants to reduce nitrates into usable forms. Some plants use it for nitrogen fixation, thus it may need to be added to some soils before seeding legumes.

Zinc - One of eight micronutrients necessary for plant health. Zinc participates in chlorophyll formation, and also activates many enzymes. Symptoms of zinc deficiency include chlorosis and stunted growth.

Micronutrients are essential elements only needed by life in small quantities. Vitamins and trace minerals are sometimes included in the term.

Microminerals or trace elements include at least iron, cobalt, chromium, copper, iodine, manganese, selenium, zinc, and molybdenum. They are dietary minerals needed by the human body in very small quantities (generally less than 100mg/day) as opposed to macrominerals which are required in larger quantities.

Vitamins are organic chemicals that a given living organism requires in trace quantities for good health, but which the organism cannot synthesize, and therefore must obtain from its diet.

Macronutrients

Macronutrients can be broken into two more groups: primary and secondary nutrients.

The primary nutrients are nitrogen (N), phosphorus (P), and potassium (K). These major nutrients usually are lacking from the soil first because plants use large amounts for their growth and survival.

The secondary nutrients are calcium (Ca), magnesium (Mg), and sulphur (S). There are usually enough of these nutrients in the soil so fertilization is not always needed. Also, large amounts of Calcium and Magnesium are added when lime is applied to acidic soils. Sulphur is usually found in sufficient amounts from the slow decomposition of soil organic matter, an important reason for not throwing out grass clippings and leaves.

Nitrogen (N)

Nitrogen is a part of all living cells and is a necessary part of all proteins, enzymes and metabolic processes involved in the synthesis and transfer of energy.

Nitrogen is a part of chlorophyll, the green pigment of the plant that is responsible for photosynthesis.

Helps plants with rapid growth, increasing seed and fruit production and improving the quality of leaf and forage crops.

Nitrogen often comes from fertilizer application and from the air (legumes get their N from the atmosphere, water or rainfall contributes very little nitrogen)

Phosphorus (P)

Like nitrogen, phosphorus (P) is an essential part of the process of photosynthesis.

Involved in the formation of all oils, sugars, starches, etc.

Helps with the transformation of solar energy into chemical energy; proper plant maturation; withstanding stress.

Effects rapid growth.

Encourages blooming and root growth.

Phosphorus often comes from fertilizer, bone meal, and superphosphate.

Potassium (K)

Potassium is absorbed by plants in larger amounts than any other mineral element except nitrogen and, in some cases, calcium.

Helps in the building of protein, photosynthesis, fruit quality and reduction of diseases.

Potassium is supplied to plants by soil minerals, organic materials, and fertilizer.

Calcium (Ca)

Calcium, an essential part of plant cell wall structure, provides for normal transport and retention of other elements as well as strength in the plant. It is also thought to counteract the effect of alkali salts and organic acids within a plant.

Magnesium (Mg)

Magnesium is part of the chlorophyll in all green plants and essential for photosynthesis. It also helps activate many plant enzymes needed for growth.

Soil minerals, organic material, fertilizers, and dolomitic limestone are sources of magnesium for plants.

Sulphur (S)

Essential plant food for production of protein.
Promotes activity and development of enzymes and vitamins.
Helps in chlorophyll formation.
Improves root growth and seed production.
Helps with vigorous plant growth and resistance to cold.

Sulphur may be supplied to the soil from rainwater. It is also added in some fertilizers as an impurity, especially the lower grade fertilizers. The use of gypsum also increases soil sulphur levels.

FOREST ECOSYSTEMS

Term of Ecosystem

A system is a group of parts that interact through one or more processes. The term ecosystem was introduced and defined by Tansley (1935), who as "a fundamental organizational unit of the natural world that includes both organisms and their spatial environment." Ecosystems have since been defined in various ways, and at different spatial and temporal scales. Some ecologists define ecosystems on the basis of biotic organisms, populations, or communities. For example, Hutchinson (1978) considered the ecosystem to be the environmental context in which population or community dynamics occur. Others define ecosystems in terms of their abiotic characteristics and processes.

For example, Lindeman (1942) defined ecosystems as "...the system composed of physical, chemical, and biological processes active within a space/ time unit." Regardless of whether the emphasis is on biotic components or abiotic characteristics and processes of ecosystems, both remain integral to the concept of ecosystem. Rowe (1961) emphasized this when he defined ecosystems as "...a three dimensional segment of the earth where life forms and the environment interact."

Wetland ecosystems have been defined in a variety of ways by researchers, resource managers, and regulatory authorities, depending on their specific needs and objectives. In the applied world of regulation, planning, and management, wetlands are usually defined in terms of their physical, chemical, and biological characteristics such as hydrologic regime, soil type, and plant species composition.

For example, in classifying wetlands for mapping, inventory, and other purposes, Cowardin *et al.* (1979) defined wetlands as "...lands transitional between terrestrial and aquatic systems where the water table is usually at or near the surface or the land is covered by shallow water..." that are characterized by the presence of hydrophytic vegetation, hydric soils, and surface water during the growing season. Wetlands are often biodiversity 'hotspots', as well as functioning as filters for pollutants from both point and non-point sources, and being important for carbon sequestration and emissions. The value of the world's wetlands are increasingly receiving due attention as they contribute to a healthy environment in many ways. Wetland functions are defined as the normal or characteristic activities that take place in wetland ecosystems or simply the things that wetlands do.

Wetlands perform a wide variety of functions in a hierarchy from simple to complex as a result of their physical, chemical, and biological attributes. For example, the reduction of nitrate to gaseous nitrogen is a relatively simple function performed by wetlands when aerobic and anaerobic conditions exist in the presence of denitrifying bacteria. Nitrogen cycling and nutrient cycling represent increasingly more complex wetland functions that involve a greater number of structural components and processes. At the highest level of this hierarchy is the maintenance of ecological integrity, the function that encompasses all of the structural components and processes in a wetland ecosystem.

Wetlands are one of the most productive of all ecosystems, and carry out critical regulatory functions of hydrological processes within watersheds. Regulating water quality, water levels, flooding regimes, and nutrient and sedimentation levels are a few of these processes. As with any natural habitat, wetlands are important in supporting species diversity and have a complex of wetland values. Moreover, the pattern of seasonal variation of the wetland affects the bird population fluctuation. Even small wetlands are extremely important to the conservation of biodiversity because they provide critical breeding habitat where dispersed populations can exchange genetic material, reducing the risks of extinction.

The present review is aimed at providing in a nutshell, the distribution of wetlands, the value of Wetlands, the causes and consequences of the loss of wetlands and their conservation status with special reference to India.

Management of Ecosystem

Practical ecosystem management usually involves maintaining a natural or artificial homeostasis in which the end product is of benefit to man, and in which violent or pathologic imbalances can be controlled. An example of a natural ecosystem which man might wish to maintain for esthetic or recreational value would be a wilderness area or National Park.

Here the most important consideration is to insure that natural processes are permitted to continue. In a forest this would mean, for example, that fire, if already a natural part of the ecosystem, be permitted to occur at natural frequency and intensity.

Several recent studies have shown that certain forests depend upon intermittent burning to recycle nutrients and minerals. In the lowland conifer forests of Alaska, if fire is excluded a thick carpet of moss may accumulate and raise the permafrost level. This encourages the growth of black spruce, a species with little food or timber value. Browsing animals such as deer, moose and bear cannot survive on black spruce.

They depend on the Tarai fires to maintain a suitable cycle of vegetation. Many foresters now feel that the underbrush and leaf litter in the Ponderosa pine forests of western United States should be removed by light ground fires every few years to maintain soil fertility, to prevent excessive accumulation of organic matter in the form of cellulose and also to prevent more devastating fires that destroy the entire forest.

In the pine forests of the South, very efficient fire suppression techniques allowed dangerous accumulation of combustible materials, and a pine fungus disease was favoured in the accumulated litter. In all of these forests, proper management should not exclude all fires, but should prevent totally devastating fires. These examples illustrate the need for understanding natural processes in ecosystems in order to maintain them properly.

The Everglades of Florida provide another example of the importance of understanding natural ecosystem processes before the system can be properly managed or maintained. The unique flora and fauna of the Everglades depends upon the proper balance of fresh and saline water flowing through its channels. If a significant decline in fresh water input occurs, as has been taking place due to the diversion of fresh water to the growing megalopolis of Miami, numerous important species of plants and animals may die and the characteristic features of the system may be destroyed. In times of drought, many animals depend upon alligator wallows as a source of water. These are pools hollowed out by alligators, and they tend to retain water longer than other parts of the swamp. If alligators decline through shooting or any other.

It is increasingly important to understand the dynamics of ecosystem, their patterns of succession and evolution, and their patterns of succession and evolution, and their patterns of diversity and stability. E.P. Odum has written a major review of these topics and has shown that ecosystem may be characterized as pioneer, developmental and mature.

Various stages within this continuum possess certain properties of production, respiration, community structure, nutrient cycling, homeostasis, etc. For example, young developmental stages of an ecosystem tend to have greater gross production in relation to respiration, but lower species diversity and less stability than mature stages. Thus, developmental stage to be favoured in agricultural practice, but they are also more vulnerable to ecologic damage or catastrophe. We will explore these relationships in more detail throughout subsequent chapters after various components of the ecosystem are considered seperately.

The Principles of Ecosystem Management

Even if some synergies between the different international arrangements and processes relating to the biodiversity of freshwater resources still deserve more attention and none of the international treaties and institutions already deals fully with freshwater ecosystems, there is an increasing consensus on the highest political levels to manage water, water-related processes, and ecosystems in a sustainable manner.

However, at the level of implementation the value of ecosystems is rarely taken fully into account. In order to close this implementation gap the contracting parties of the CBD and other advocates have proposed a so-called "ecosystem-based approach" as an appropriate way of managing natural resources.

The Ecosystem Approach

The ecosystem approach is a strategy for the integrated management of land, water and living resources that promotes conservation and sustainable use in an equitable way. The term ecosystem is used more as a mental construct suggesting complexity and systems interaction rather than a geographic entity. An emphasis on the complexity of system-wide interactions highlights scientific uncertainty and results in the need to deal with this uncertainty explicitly by acting conservatively and managing adaptively: setting a course of action based on a set of hypotheses, monitoring what happens, and re-evaluating the direction on what one learns. At its most fundamental level, an ecosystem approach maintains diversity as a means of building resilience against catastrophic events in biological, economic, organisational, and political systems. At its second meeting in 1995, the Conference of the Parties of the CBD adopted the ecosystem approach as the primary framework for action, and subsequently has referred to the ecosystem approach in the elaboration and implementation of the various thematic and cross-cutting issues work programmes under the Convention. The issues concerned include:

- Biological diversity of inland water ecosystems.
- Marine and coastal biological diversity.
- Agricultural biological diversity.
- Forest biological diversity.
- Indicators of biological diversity.
- Incentive measures and environmental impact assessment.

By recommending the ecosystem approach as the guiding concept, in May 2000 the fifth COP took a next step towards its practical verification in the frame of the CBD.

Concerning inland water ecosystems a work programme was adopted by the COP which highlights the importance of designing integrated watershed, catchment and river basin management strategies. Particularly the interrelation between water management and agriculture was stressed, as well as the need to use the ecosystem approach in all sectors which have an impact on inland water biodiversity.

The ecosystem approach was initially developed as a research paradigm and not intended to serve as the basis for resource management. With the political application of the notion "ecosystem-based approach" advocates intended to find an adequate language in response to the shortcomings of classical nature conservation approaches.

Although classical nature conservation (*e.g.* designation of protected areas) can be considered successful in many regions and also is an important module in ecosystem management, its recognition of human needs and other legitimate sectoral interests was rather insufficient. It did not put enough emphasis on the biological diversity outside protected areas. Furthermore, the primary method of protecting freshwater biodiversity has been to designate particular species as threatened or endangered making them subject to national recovery programmes or international protection.

Unfortunately, this approach is failing and it demonstrates the need for a more holistic approach in order to overcome the shortcomings of traditional efforts in nature protection policy. In the United States, for example, no aquatic species has ever graduated from the government's endangered species list, but 10 species of fish have been removed due to extinction. Equally, ecosystem management responds to the obvious shortcomings of classical single-resource approaches focused on exploitation.

The 12 Malawi Principles

- Management objectives are a matter of societal choice.
- Management should be decentralised to the lowest appropriate level.
- Ecosystem managers should consider the effects (actual or potential) of their activities on adjacent and other ecosystems.
- Recognising potential gains from management there is a need to understand the ecosystem in an economic context. Any ecosystem management programme should:
  - Reduce those market distortions that adversely affect biological diversity.
  - Align incentives to promote sustainable use.
  - Internalise costs and benefits in the given ecosystem to the extent feasible.
- A key feature of the ecosystem approach includes conservation of ecosystem structure and functioning.

- Ecosystems must be managed within the limits to their functioning.
- The ecosystem approach should be undertaken at the appropriate scale.
- Recognising the varying temporal scales and lag effects which characterise ecosystem processes, objectives for ecosystem management should be set for the long term.
- Management must recognise that change is inevitable.
- The ecosystem approach should seek the appropriate balance between conservation and use of biological diversity.
- The ecosystem approach should consider all forms of relevant information, including scientific and indigenous and local knowledge, innovations and practices.
- The ecosystem approach should involve all relevant sectors of society and scientific disciplines. These principles were agreed at CBD's COP-5 under decision V/6 with minor modifications.

Therefore, ecosystem management is proposed as a modern way of managing natural systems and generally intends to use natural resources within their natural limits. However, although there are still some controversies on how to translate ecosystem management into concrete management terms, parties of the CBD have already agreed on basic principles to guide the *modus operandi*.

A key point is that the ecosystem approach is a holistic process for integrating and delivering in a balanced way the three objectives of the CBD: conservation and sustainable use of biological diversity combined with equitable sharing of biodiversity's benefits. In particular, the agreed principles of the ecosystem approach stress the importance of:

- Societal choice concerning management objectives.
- The "economics" of biodiversity (internalisation of costs and benefits, balance between conservation and use).
- Adequate institutional structures (decentralisation, management at the appropriate scale).
- Adaptive management which enables policy makers to anticipate and cater for the natural dynamics of ecosystems.
- Consideration not only of scientific knowledge but also of indigenous and local knowledge.

In order to translate this rather abstract principles into guidelines for water management it is first necessary to define the appropriate management level to meet the general objectives.

While the cited principles stress the importance of decentralised approaches in general terms, the idea of river basin management is considered an adequate starting point for the implementation of the ecosystem approach in water management.

The reasoning is that river basins offer many advantages for strategic planning and that they represent the most appropriate geomorphologic units on which to base ecosystem management practices.

However, concerning ecosystem protection there are some limits: groundwater catchments only rarely match with surface water catchments; water quality and quantity inside the river basin interacts with other influences outside the catchments boundaries; estuarine ecosystems, for example, can only be sufficiently protected by taking into account the interaction of freshwater and tidal currents.

Therefore, river basin management should not be perceived as a panacea but it does provide a sound geographical basis for implementation. Briefly, ecosystem management should be an integral component of comprehensive river basin management strategies. Hence, the explicit switch to ecosystem management leads to an accentuation of the main focus in water management because ecosystem management demands:

- A broader perspective: whole ecosystems and the full range of services, no sectoral view.
- Efforts to balance natural ecosystem robustness and human-induced alterations.
- Collation of information on the status of, and threats to, the biodiversity of the freshwater ecosystem concerned (including impacts of the introduction of alien species).
- Explicit consideration of land-use, in-stream flow and water temperature for maintaining ecological functions.
- A long-term perspective.
- Consideration of the full economic value of freshwater ecosystems (include its biodiversity).
- Recognition of ecosystem's change as a matter of fact (*i.e.* restoration of damaged or destroyed ecosystems will never lead to the "pristine" state; adaptive management to deal with uncertainties).
- Integration of scientific knowledge and modern management tools (Decision Support Systems, Geographical Information Systems etc.) into decision-making.
- Identification of win-win options between protection and use of freshwater ecosystems: "use them, but don't lose them".
- Consideration of the full range of policy instruments applicable in water management (economic and legal instrument, co-operation etc.)

As most of these elements are widely agreed upon the rhetoric of international water policy, the next steps must include attempts to tackle the implications of the ecosystem approach in practice. For example, only little is known about the value of ecosystem management in political terms for at least two reasons:

- The concept appears highly demanding because much information – sometimes on "invisible" relationships between human-induced pressures and natural reactions – is needed, and effective vertical and horizontal co-ordination of sector policies and planning instruments has to take place. Today, even in wealthy nations a lack of integration across sectors in managing water resources can be identified.
- It is necessary to decide which services of water receive priority, and how the costs and benefits are to be distributed. The developed world, for example, can "afford" to stop or to remove development projects in favour of ecosystem integrity (*e.g.* Great Whale Hydropower Project in Canada, Edwards Dam in the US) and to use sometimes generous subsidies to compensate land owners for restricting their use of the lands. The restoration of river ecosystems in the developed world also is an example for costs emerging in the short terms. Although long-term economic benefits are regularly provided, political decision are often dominated by short-term considerations.

Water and Forested Ecosystems

Ecologists consider water to be the defining part in an ecosystem, including the forest ecosystem. Water shapes the physical landscape through erosion Undulating forest surrounds meadows, lakes, and rivers of the Charlevoix Conservation Area, a coastal watershed in Quebec, Canada. This region, which illustrates a high-quality forested watershed, has been designated a biosphere by the United Nations Educational, Scientific and Cultural Organization (UNESCO). and deposition. It also shapes the biological parts of the ecosystem by its presence or absence; its quantity and quality; and its occurrence and distribution.

The water cycle plays a key role in ecosystem functions and processes. Forests, in turn, are vital to the water cycle and to water quality. In essence, the forest acts like a giant sponge, filtering and recycling water. Approximately 80 per cent of U.S. fresh-water resources are estimated to originate in forests, which cover one-third of the U.S. land area.

Tree leaves intercept water from rain, snow, and fog; the leaves also release water back to the atmosphere by evapotranspiration. Tree roots extract water from the soil while helping hold the soil in place. Forested land reduces the surface impact of falling rain through interception and delay of water reaching the surface. Forestland also decreases the amount and velocity of storm runoff over the land surface.

This in turn increases the amount of water that soaks into the ground, a portion of which can ultimately recharge underlying aquifers. Conversely, water from hydraulically connected surficial aquifers may enter streams and wetlands, helping to maintain their water levels during dry periods.

Forests and the Hydrologic Cycle

The surface water in a stream, lake, or wetland is most commonly precipitation that has run off the land or flowed through topsoils to subsequently enter the waterbody. If a surficial aquifer is present and hydraulically connected to a surface-water body, the aquifer can sustain surface flow by releasing water to it. In general, a heavy rainfall causes a temporary and relatively rapid increase in streamflow due to surface runoff. This increased flow is followed by a relatively slow decline back to baseflow, which is the amount of streamflow derived largely or entirely from groundwater.

During long dry spells, streams with a baseflow component will keep flowing, whereas streams relying totally on precipitation will cease flowing. Generally speaking, a natural, expansive forest environment can enhance and sustain relationships in the water cycle because there are less human modifications to interfere with its components. A forested watershed helps moderate storm flows by increasing infiltration and reducing overland runoff. Further, a forest helps sustain streamflow by reducing evaporation (*e.g.*, owing to slightly lower temperatures in shaded areas). Forests can help increase recharge to aquifers by allowing more precipitation to infiltrate the soil, as opposed to rapidly running off the land to a downslope area.

Riparian Areas

The riparian zone is broadly defined as the area between a body of water and the upland parts of the landscape that are rarely flooded except under the most extreme conditions.

But the term also can refer more specifically to the immediate streamside area. Riparian areas represent less than 10 per cent of most forest ecosystems, yet these areas often are the most productive portions. Compared to upland regions, riparian areas have more water available; the vegetation is more robust; the soils are deeper; the timber often is of higher quality; and the waterbodies have more shade.

The riparian zone also may include wetlands bordering streams and lakes. This combination of factors makes riparian areas among the most heavily used portions of a forest. Riparian and wetland areas provide abundant and reliable forage for wildlife, as well as transportation corridors. They also may receive heavy human use for recreation.

Riparian zones also are attractive destinations for logging and for livestock grazing; as a result, riparian areas in forests are sometimes heavily damaged, especially in the forests of the arid American Southwest. Fortunately, riparian areas respond well to good management practices.

Aquatic Biodiversity

Forest lands and waters are vitally important in maintaining biodiversity and providing habitat for fish and wildlife, including threatened or endangered aquatic species. In the United States, over one-third of national forest lands are critical for maintaining aquatic biodiversity and protection of listed species. For aquatic species, watersheds provide the basic unit of any conservation strategy. Many watersheds also contain isolated habitats with unique characteristics producing a high potential for rare species. Some species occur only near a single spring or in a single stream within a given watershed. Lands set aside to protect these unique habitats also benefit the entire watershed and its ecosystem.

Development of Community Wastelands and Denuded Forests

Conservation of biodiverstity and protection of forests are the other important components of community forestry. However there has been a lukewarm response to these activities in the past, mainly because of lack of awareness, inadequate benefits and lack of trust between the local communities and government agencies who own these lands. Hence the new strategy should find a suitable solution for these problems.

Development of forest lands: Rehabilitation of denuded forest lands and protection of natural forests are essential to conserve biodiversity. The forests influence the climate and soil moisture conditions and improve agricultural production.

This aspect needs wider publicity in rural areas and the local people should be motivated to take active part in Joint Forest Management. The Village Forest Committees willing to protect the natural forests, should be encouraged to share the benefits in the form of fodder, fuel, timber and NWFP. These committees can also take up reforestation of degraded forest lands, aiming to conserve biodiversity and generate income. While facilitating natural regeneration, the local communities can also introduce fuel, timber and NWFP species of their choice to enhance the productivity on denuded forest lands.

Development of community lands: Lack of people's initiative was a serious problem in developing community wastelands, mainly due to lack of direct benefits. Earlier efforts to establish woodlots could not generate substantial income. Establishment of NWFP species such as tamarind (*Tamarindus indica*), Indian gooseberry (*Emblica officinalis*), neem (*Azadirachta indica*), Pongamia (*Derris indica*), *Madhuca spp*. and a wide range of medicinal species can generate substantial income, regularly for over 50-100 years. With the world's herbal medicine business likely to cross US $ 3 trillion during the next decade, it will be an excellent opportunity for the Asian farmers to enhance their income through cultivation of these species on community lands as well as on private wastelands.

Development of Mined Areas: Rehabilitation of mined areas is another cause of serious concern. Although the industries are obliged to establish a green cover after mining the area, the outcome has been very poor. Such sites are major causes of soil erosion and flooding of rivers. Lack of commitment, inadequate expertise in selection of plant species and their management have been the main reasons for failure.

These activities can be entrusted to voluntary organisations and local communities, with research support to raise healthy plantations. Preference should be given to NWFP and timber species which have long gestation period, sufficient to enrich the mined lands.

CO_2 Sinks: There are opportunities for environmental protection by establishing industrial greenbelts, recycling treated effluent and developing carbon dioxide sinks, particularly around urban areas. The polluting industries can be compelled to establish greenery through the involvement of voluntary organisations and local people.

Environmental Awareness and Eco-tourism: Lack of awareness among the common public is a cause for environmental pollution in the developing countries. Hence, continuous efforts are needed to motivate the common public to plant more trees and protect the environment. School children and literate urban population are fairly receptive to participate in environmental protection. This can be undertaken through school based plant nurseries, development of arboretum, botanical gardens and greenbelts around the schools and industries.

While developing industrial greenbelts and afforestation on river banks, special efforts can be made to promote eco-tourism. Creating waterbodies through plugging of gullies, developing, nature trails, shady grounds for resting, selecting suitable plant species to attract wildlife and providing adequate eco-tourism. Environmental exhibitions at these sites can help in motivating the public to take active part in community forestry and environmental protection.

Organisational Development

Development of local organisations is the key to ensure people's participation and to sustain the benefits of investments and efforts made in community forestry.

Tree Growers Organisations

The initial need is to provide leadership to organise local people in afforestation. This can be carried out through formation of local Selp-Help Groups, village level farmers' associations and forest groups. This activity should receive priority during the next decade.

Capacity Building

Training of local farmers, supply of critical inputs, credit and technical advisory services to improve the profitability and developing market networks are the critical inputs which have to be introduced in the field to support tree growers.

Networking of Tree Growers and User Groups

There is an urgent need to set up networks of tree growers, forestry users and concerned scientists in these fields. Such networks should identify the opportunities for tree growers. They can also promote processing and marketing of forestry products and protect the interests of tree growers.

Forestry Information Service

Setting up of information centres to identify the priority of the industries and farmers and provide necessary information to enhance the economic viability of community forestry would help to sustain the interest of tree growers. Forestry Extension Units should closely associate with Information Service Centres to identify appropriate technologies for transferring to the field.

Forestry Research

The networks should identify the problems affecting the production and marketing of the produce and encourage the Forestry Research Institutions to take up relevant scientific studies to solve the problems. The researchers should closely associate with the network and Information Centres to disseminate new technologies to solve field problems and enhance the production.

Forestry research should focus on the following topics:

- Conservation of biodiversity;
- Improving the productivity of wastelands, tree plantations and natural forests;
- Exploring the utility of multipurpose tree species and NWFP;
- Developing of alternatives for conserving wood;
- Selecting various organisational systems to manage community forestry.

The outcome of these research topics can further strengthen the community forestry programme in the country.

Invertebrate Species in Forest Ecosystems

The diversity and abundance of invertebrate species in forest ecosystems are determined by the interaction of complex ecological and environmental factors. Research into these factors is gradually providing data that can be incorporated into models, both descriptive and mathematical, of the processes involved.

At the core of such models is the description of the availability of resources that enables a particular invertebrate species to reproduce, whereas the rate of reproduction is dependent on multiple factors, including temperature, competition, etc., that determine utilization of available resources. In relation to exploitation of resources within a forest ecosystem, the primary focus for forest managers is the potential of invertebrates to cause damage and to reduce yields of the trees themselves.

This is certainly the case in managed forests where the primary aim is to grow trees for commercial gain, often with associated value from recreational, landscape or environmental considerations. Any invertebrates that feed directly on the trees, or have an influence of how the trees grow, may be classified as pests and hence may require some form of pest management to reduce the threat.

But it is pertinent to consider how ecological theory and practice can be used both to understand why a pest problem has arisen and to provide guidance on how to combat that problem. The same principles can be applied to understanding and enhancing the biodiversity of forests, including conservation of rare and beneficial invertebrates. Indeed, the two extremes of pest outbreak and invertebrate rarity are part of the same continuum of interaction between invertebrates and their host trees.

In general, considering that trees are known to harbour the greatest numbers of insect species (diversity), it is surprising that serious pest status is reached only relatively infrequently.

For example, although oak *(Quercus* spp.) is known to support at least 450 species of insect in Britain , Bevan (1987) lists only 35 species as pests, of which five were classified as important or severe (XXX-XXXXX on Bevan's scale).

Wider listing of pest species has been noted for forest insects in continental Europe, although even here the number of consistently serious pests is small relative to the total diversity of insects present (Klimetzek 1993).

Naturally, the severity of a pest outbreak will be classified differently depending on the purposes of the forest manager, and thus it is not always possible to classify insects consistently as pests across geographical ranges or, indeed, between years. Further, an insect may normally be classified as uncommon or rare but, given an abundance of resources, may be described later as a pest.

This was certainly the case for oak pinhole borer, *Platypus cylindrus* (Coleoptera: Platypodidae), which prior to the 1987 great storm in southern England was listed in the British Red Data Book as rare.

Winter (1988) warned that the great increase in availability of freshly dead oak following the storm could result in an increase in numbers of *P. cylindrus* and, subsequently, it has been noted that the insect is now causing damage to high-value oak logs as a result of its habit of boring directly into the heartwood. Therefore, consideration of why invertebrate abundance might increase above a given threshold level can provide a more consistent appraisal of status and also potential approaches for reducing populations below that threshold or indicating measures that can be taken to encourage rare or endangered species.

Among the many interacting processes that determine insect diversity and abundance, a few attributes that are particularly well linked to the environmental status of trees can be distinguished. These provide a means to describe intrinsic processes but also allow comparison with other plants, thus offering the potential for a more consistent approach to descriptive population ecology of invertebrates irrespective of the host plant being considered.

Longevity

Trees are the longest-lived components of an ecosystem and, in the majority of natural situations, approach or represent the climax stage of vegetation succession. This provides opportunities for continuity of insect population growth and over time this will give a particular set of insect associates for a given tree species.

Continuing the theme of resource availability, the fact that trees may be present in the same location for tens or hundreds of years ensures that both migrant and, particularly, relatively sedentary invertebrates can exploit them at some stage during their growth.

However, longevity *per se* is not a quantifiable attribute but does provide the means for the other components, to influence invertebrate diversity and abundance. Increasing stability in the environment, as represented under forest conditions, leads to stability in both microclimate and resource availability, particularly favouring exploitation by specialist insect feeders and their associates.

Biogeography

Tree species richness in a given region or locality varies considerably with geographical location and thus the climax community structure and longevity of individual species may vary accordingly. Although subject to uncertainty and lack of unequivocal evidence of the underlying processes, there is certainly a gradient of species richness of plant cover, especially trees, from the tropics to the poles.

Not surprisingly, the same applies to a wide range of animals and plants; as an extreme example, the species richness of orchids ranges from 15 to 2500 in a latitudinal range from 0 to 66° S.

Further examples are provided by Brown (1988). More specifically, in an analysis of species richness of deciduous forests both in the tropics and, latitudinally, within North America, Brown and Gibson (1983) contrasted the high diversity of tree species in the tropics (up to 100 ha^{-1}) with the low diversity from north to south within North America (1–30ha^{-1}). However, even within such a generalization there are exceptions, so that for example conifers are more diverse in temperate latitudes.

Geological Time

The ecological structure of forests and of their associated invertebrate faunas is determined by both present and recent characteristics, particularly area, and by their history in geological time. The latter characteristic is driven as much by *time per se* as by the history of major disturbances (glaciations, volcanic eruptions, land shifts, etc.) within geological time. Thus, present flora and fauna may well be determined by recolonization since the last major disturbance, the diversity of species being a reflection of both colonization and speciation relative to more recent ecological conditions.

The gradient of species from polar to tropical regions could therefore be explained by the high level of disturbance closer to the polar masses compared with the relative stability of climate in the tropics. Evidence from habitats that have become isolated since the last major disturbance indicates that current species diversity may reflect the historical link to the original major habitat. For example, islands that were once joined by land bridges to the larger land masses of New Guinea have species diversities of birds greater than expected compared with other islands that were never connected. However, although local ecological communities can be explained by historical patterns, the gradients over wide geographical scales are not so easily attributed to time as the major determinant.

The concept that major disturbances drive change undoubtedly carries weight, but closer analysis of tropical rain forests reveals that they too have been subject to disturbances of various sorts and are, in many cases, much changed over geological time. Based on this knowledge Haffer proposed a theory that disturbances in the tropics during the Pleistocene (alternating wet and dry periods giving rise to forest blocks surrounded by grassland) gave rise to refugia in which high levels of speciation took place, ultimately leading to the current high diversity characteristic of the tropics. While this theory is plausible and has been invoked by other authors, tests of the hypothesis have cast doubt over its wide applicability.

Irrespective of the mechanistic bases for variation in species richness over wide latitudinal ranges, time is a significant determinant at a more local scale. Birks (1980) analysed the insect faunas of trees in Britain during the period since the last glaciation 13000 years ago and concluded that there was a significant correlation with the time that a given tree species had been continuously present in Britain during that period.

Taking this further, Kennedy and Southwood (1984) included Birks' radiocarbon estimates of time in Britain as a variable in a multiple regression analysis of insects on British trees.

They showed that time was a significant predictor, both in its own right and as a co-predictor with log abundance (area). This is logical considering that over time there will be a tendency for well-established tree genera to increase their abundance, despite the recent proliferation in areas arising from commercial afforestation.

Size

A consequence of basic structure and longevity is that trees tend to be the largest components of the phylloplane, thus providing a much wider number and range of ecological niches compared with herb and shrub layer plants. Such attributes enable more insect species to colonize the plants without facing competition, both interspecific and intraspecific. Size has many interrelated characteristics that help to explain the presence of both herbivores and their natural enemies.

- A larger physical presence is more easily detected and colonized by invertebrates. This concept, termed 'apparency', reflects a higher visual profile for colonizing adults or passively migrating immature life stages (*e.g.* mean height was a good predictor in models of the distribution of Scolytidae in Finland; Heliovaara & Vaisanen 1995) and a greater production of olfactory cues.
- Complex plant architecture is a consequence of greater size and for trees is particularly linked to the provision of many niches capable of supporting invertebrates with a wide range of feeding strategies (guilds). Although the concept of the guild is not always clearly defined , it is nevertheless a useful means of describing the trophic interactions of invertebrates and trees without having to consider their precise taxonomic status. Thus, seed, bud, leaf, bark and wood feeders can be distinguished, each occupying a different part of the tree and therefore potentially avoiding competition.
- Increased biomass provides greater food resources per individual, thus encouraging growth and breeding success whilst reducing potential intraspecific competition. Price (1992) has reviewed the role of plant resources in insect population dynamics and concludes that, in part, the carrying capacity of an ecosystem depends on plant succession in time and space. Carrying capacity is therefore greater for trees, which generally appear late in succession, compared with herbs or shrubs. Thus biomass is available for longer and in greater quantities on trees compared with other plants.
- The concept of enemy-free space applies to the likelihood of herbivores being overlooked by searching parasitoids or predators. One aspect of size is that there is more opportunity for herbivores to occupy a part of the plant that offers protection against natural enemies (and competition from other herbivores). The complexity and number of niches present on a plant (plant architecture; Lawton 1983) therefore has an effect on enemy-free space. The greater the

architectural complexity, the greater the time that natural enemies may have to spend in searching for prey, thus reducing their overall effectiveness and enhancing the survival probability of herbivores.

Area

Forests tend to occupy greater areas than other plants. There is evidence to support the view that the greater the area occupied by the host plant, the greater the number of insect species associated with that host. This is called the species-area relationship and was originally linked to the island biogeography theories of species isolation and colonization.

Species-area relationships have been demonstrated for many plants and animals, including both highly mobile and sedentary species. With regard to the numbers of insect species on trees, data on the British insect fauna, one of the best characterized in the world, have been analysed by a number of authors.

Arising from earlier tentative analyses , Kennedy and Southwood (1984) analysed the relationships between insect species and characteristics of tree species in Britain. They concluded that there was a significant relationship between the area occupied by British tree species and the numbers of insect species associated with them. The area occupancy data were based on county and subcounty information of plant records and were thus potentially overestimates of actual areas occupied.

Claridge and Evans (1990) reanalysed the species-area relationships using the same insect data as Kennedy and Southwood but employing actual area information derived from Forestry Commission census data. They concluded that area was not a good descriptor for insect species on deciduous trees but was just significant for conifers.

However, further unpublished analysis of the data broken down into numbers and areas of habitat patches occupied by the tree genera indicated that whereas total area was not a good descriptor, the degree of fragmentation of that area provided a better explanation for the observed insect species numbers. A similar conclusion, concentrating on Lepidoptera that have moved hosts to include conifers, was reached by Fraser and Lawton (1994) in an analysis of new associations on British trees. The overall finding that area provides a broad descriptor of insect diversity remains valid, although this does not explain the mechanisms that lead to these associations.

Taxonomic Relatedness

In a given ecosystem, the closer the taxonomic relatedness of host plants, the more similar the numbers of insect species associated with those plants. This has been documented for tree species that have a long history in a particular location and thus provide rich sources of invertebrate species that may subsequently colonize other tree species, especially if they are taxonomically related. A particularly good example of the process in action is the colonization of *Nothofagus,* a member of the Fagaceae newly introduced to Britain, which has been colonized rapidly by insects associated with closely related members of the native Fagaceae.

Geographical isolation as a critical factor in determining invertebrate diversity and implications for international movement of pest organisms. Geographic isolation drives the acceleration of speciation and provides the means to derive principles on the processes involved. In addition, the fact of isolation means that some invertebrates abundant in one region will be absent elsewhere, despite conditions being suitable for their establishment. In such situations, it is pertinent to examine the international plant health implications of geographical isolation and how modern patterns of trade influence the degree of isolation and later colonization by new insect species.

Geographical Isolation as a Factor in Determining Insect Diversity

In general, the basic principles apply across ecosystem boundaries and geographical zones extending from the poles to the tropics. In addition, the isolation afforded by relatively small islands provides further evidence of the principles of species colonization over space and time. Great Britain provides a particularly good example, because of its island position in the Atlantic Offshore climatic zone and the fact that it has been subjected to glaciation in recent geological time.

Among the ecological characteristics that arise from this set of circumstances are a relatively depauperate mix of both animal and plant species that have arisen from the combination of losses resulting from the last ice age and the slow rate of recolonization arising from island status. British flora and fauna are therefore impoverished relative to continental Europe. Further influences arise from the preponderance of exotic tree species in commercial forestry in Britain, which although dominated by exotic conifer species also has increasing numbers of exotic broadleaved species.

Expanding on the principle that colonization of tree species is determined by the features, the ranges of numbers of insect species associated with trees in Britain reflect both the area (with provisos) and length of time that trees have been continuously present in Britain. However, there are exceptions, so that Sitka spruce in particular would appear to support far fewer species than predicted on the basis of area, although this probably also reflects the relatively short history of the species and recent expansions in areas planted in Britain.

Implications for International Movement of Pest Insects

Trading patterns between countries have had a considerable influence on both the flora and fauna of most of the countries in the world. This reflects deliberate introductions of plants and animals for agricultural, forestry and amenity purposes and also accidental introductions by association with other commodities. In the majority of cases, the introduced species have had a neutral to beneficial effect, reflecting their 'domesticity' within managed crop or animal husbandry systems.

However, there are many exceptions to this generalization and in most cases the results of introduction of new species were not foreseen, either through lack of knowledge of the potential impacts or because the introduction was accidental and undetected.

The literature on international biological control is dominated by examples where attempts have been made to rectify the effects of new pest introductions (both animal and plant).

A classic example is the control of the woodwasp *Sirex noctilio* (Hymenoptera: Siricidae) in exotic pine forests in Australia with the introduced nematode *Deladenus siricidicola* (Nematoda: Neotylenchidae) isolated from *Sirex* in New Zealand. This nematode has a life cycle that includes a free-living form that feeds on the fungus *Amylostereum areolatum* and a parasitic form that invades *S. noctilio* larvae and subsequently migrates to the reproductive organs of the female wasp, causing sterility. In this situation, all elements of the interaction are exotic to the new location, with little prospect of natural control being realized.

ensuring that the resource is conserved to meet the needs of future generations. FAO works to improve the knowledge on sustainable forest and wildlife management, and supports the development and implementation of appropriate policies and practices to ensure forest and wildlife protection in order to maintain or improve their capacity to produce wood and non-wood products, sustain wildlife populations, conserve biodiversity, safeguard wildlife habitat, mitigate climate change, and protect soils and watersheds.

What is Biodiversity?

More than 10 million different species of animals, plants, fungi and micro-organisms inhabit the Earth. They and the habitats in which they live represent the world's biological diversity, or biodiversity as it is often called. Humans use at least 40,000 species of plants and animals on a daily basis for food, shelter, clothing and medicinal needs.

Much of the global concern with deforestation focuses on the alarming loss of biodiversity. There is also considerable concern with the poverty of many forest dependent communities. Many poor communities around the world rely on local biodiversity for a range of essential services. These include materials for housing and clothing, food from a range of wildlife species and traditional medicines derived from local plants and animals.

The populations of developed nations also depend on biodiversity for their survival and quality of life. Close to 40 percent of the pharmaceuticals used in the United States are either based on, or synthesized from, natural compounds found in plants, animals or microorganisms.

The greatest value of biodiversity might still be unknown. Only a fraction of known species has been examined for potential medicinal, agricultural or industrial value. Nor do we fully understand how biodiversity contributes to the well-being of the larger global environment. And we are only just beginning to learn how biodiversity helps communities around the world satisfy their economic, dietary, health and cultural needs.

One thing is certain: the more we learn about biodiversity the more we realize how much the world depends on it. Yet whole species of plants, animals, fungi, and microscopic organisms are being lost at alarming rates.

Forest Biodiversity

Forests are the most diverse ecosystems on land, because they hold the vast majority of the world's terrestrial species. Some rain forests are among the oldest ecosystems on Earth. Timber, pulpwood, firewood, fodder, meat, cash crops, fish and medicinal plants from the forest provide livelihoods for hundreds of millions of people worldwide. But only a fraction of known species has been examined for potential medicinal, agricultural or industrial value.

A Continuing Threat

Forest biodiversity is threatened by rapid deforestation, forest fragmentation and degradation, hunting and the arrival of invasive species from other habitats. We are losing 12 million hectares of forest a year, much of it tropical rainforest with its unique and rich biodiversity.

How can we Protect Biodiversity?

One of the best ways to conserve forest biodiversity is to establish protected forest areas. But these areas must be of a certain size, or consist of a well-designed network of forest areas, to allow the local forest ecosystems to continue operating effectively.

The forest surrounding the protected area must then be carefully managed so that it serves as a buffer zone. These surrounding forests also allow local communities to earn a livelihood without infringing on the protected forest.

There have been numerous efforts aimed at safeguarding the world's biodiversity by protecting species in areas outside their original habitats. For example, seeds of some of the most economically important trees are being conserved in seed centers and gene-banks as a way of protecting their genetic diversity.

But a large number of forest species have seed that do not survive storage, and many species of animals and plant-life are hard to protect once removed from their ecosystems.

CIFOR's Focus

Biodiversity will continue to change and it is impossible to protect everything, so we need to decide which species are critically important and how they can best be maintained. The challenge is to ensure that we do not focus only on the needs of developed nations and the world's economic elite. We must also recognize and examine the priorities of local people who depend on forests, especially in developing nations.

CIFOR is meeting this challenge by studying forest biodiversity and increasing the understanding of this unique resource in ways that accommodate both local and global forest values. It is also researching logging techniques that have a reduced impact on forests and biodiversity while still remaining profitable for logging companies.

INVERTEBRATE SPECIES IN FOREST ECOSYSTEMS

The diversity and abundance of invertebrate species in forest ecosystems are determined by the interaction of complex ecological and environmental factors. Research into these factors is gradually providing data that can be incorporated into models, both descriptive and mathematical, of the processes involved.

At the core of such models is the description of the availability of resources that enables a particular invertebrate species to reproduce, whereas the rate of reproduction is dependent on multiple factors, including temperature, competition, etc., that determine utilization of available resources. In relation to exploitation of resources within a forest ecosystem, the primary focus for forest managers is the potential of invertebrates to cause damage and to reduce yields of the trees themselves.

This is certainly the case in managed forests where the primary aim is to grow trees for commercial gain, often with associated value from recreational, landscape or environmental considerations. Any invertebrates that feed directly on the trees, or have an influence of how the trees grow, may be classified as pests and hence may require some form of pest management to reduce the threat.

But it is pertinent to consider how ecological theory and practice can be used both to understand why a pest problem has arisen and to provide guidance on how to combat that problem.

The same principles can be applied to understanding and enhancing the biodiversity of forests, including conservation of rare and beneficial invertebrates. Indeed, the two extremes of pest outbreak and invertebrate rarity are part of the same continuum of interaction between invertebrates and their host trees. In general, considering that trees are known to harbour the greatest numbers of insect species (diversity), it is surprising that serious pest status is reached only relatively infrequently.

For example, although oak *(Quercus* spp.) is known to support at least 450 species of insect in Britain, Bevan (1987) lists only 35 species as pests, of which five were classified as important or severe (XXX-XXXXX on Bevan's scale). Wider listing of pest species has been noted for forest insects in continental Europe, although even here the number of consistently serious pests is small relative to the total diversity of insects present (Klimetzek 1993).

Naturally, the severity of a pest outbreak will be classified differently depending on the purposes of the forest manager, and thus it is not always possible to classify insects consistently as pests across geographical ranges or, indeed, between years. Further, an insect may normally be classified as uncommon or rare but, given an abundance of resources, may be described later as a pest. This was certainly the case for oak pinhole borer, *Platypus cylindrus* (Coleoptera: Platypodidae), which prior to the 1987 great storm in southern England was listed in the British Red Data Book as rare.

Winter (1988) warned that the great increase in availability of freshly dead oak following the storm could result in an increase in numbers of *P. cylindrus* and, subsequently, it has been noted that the insect is now causing damage to high-value oak logs as a result of its habit of boring directly into the heartwood. Therefore, consideration of why invertebrate abundance might increase above a given threshold level can provide a more consistent appraisal of status and also potential approaches for reducing populations below that threshold or indicating measures that can be taken to encourage rare or endangered species. Among the many interacting processes that determine insect diversity and abundance, a few attributes that are particularly well linked to the environmental status of trees can be distinguished. These provide a means to describe intrinsic processes but also allow comparison with other plants, thus offering the potential for a more consistent approach to descriptive population ecology of invertebrates irrespective of the host plant being considered.

LONGEVITY

Trees are the longest-lived components of an ecosystem and, in the majority of natural situations, approach or represent the climax stage of vegetation succession. This provides opportunities for continuity of insect population growth and over time this will give a particular set of insect associates for a given tree species. Continuing the theme of resource availability, the fact that trees may be present in the same location for tens or hundreds of years ensures that both migrant and, particularly, relatively sedentary invertebrates can exploit them at some stage during their growth.

However, longevity *per se* is not a quantifiable attribute but does provide the means for the other components, to influence invertebrate diversity and abundance. Increasing stability in the environment, as represented under forest conditions, leads to stability in both microclimate and resource availability, particularly favouring exploitation by specialist insect feeders and their associates.

BIOGEOGRAPHY

Tree species richness in a given region or locality varies considerably with geographical location and thus the climax community structure and longevity of individual species may vary accordingly. Although subject to uncertainty and lack of unequivocal evidence of the underlying processes, there is certainly a gradient of species richness of plant cover, especially trees, from the tropics to the poles. Not surprisingly, the same applies to a wide range of animals and plants; as an extreme example, the species richness of orchids ranges from 15 to 2500 in a latitudinal range from 0 to 66° S. Further examples are provided by Brown (1988).

More specifically, in an analysis of species richness of deciduous forests both in the tropics and, latitudinally, within North America, Brown and Gibson (1983) contrasted the high diversity of tree species in the tropics (up to 100 ha^{-1}) with the low diversity from north to south within North America (1–30ha^{-1}). However, even within such a generalization there are exceptions, so that for example conifers are more diverse in temperate latitudes.

GEOLOGICAL TIME

The ecological structure of forests and of their associated invertebrate faunas is determined by both present and recent characteristics, particularly area, and by their history in geological time. The latter characteristic is driven as much by *time per se* as by the history of major disturbances (glaciations, volcanic eruptions, land shifts, etc.) within geological time.

Thus, present flora and fauna may well be determined by recolonization since the last major disturbance, the diversity of species being a reflection of both colonization and speciation relative to more recent ecological conditions. The gradient of species from polar to tropical regions could therefore be explained by the high level of disturbance closer to the polar masses compared with the relative stability of climate in the tropics. Evidence from habitats that have become isolated since the last major disturbance indicates that current species diversity may reflect the historical link to the original major habitat.

For example, islands that were once joined by land bridges to the larger land masses of New Guinea have species diversities of birds greater than expected compared with other islands that were never connected. However, although local ecological communities can be explained by historical patterns, the gradients over wide geographical scales are not so easily attributed to time as the major determinant.

The concept that major disturbances drive change undoubtedly carries weight, but closer analysis of tropical rain forests reveals that they too have been subject to disturbances of various sorts and are, in many cases, much changed over geological time.

Based on this knowledge Haffer proposed a theory that disturbances in the tropics during the Pleistocene (alternating wet and dry periods giving rise to forest blocks surrounded by grassland) gave rise to refugia in which high levels of speciation took place, ultimately leading to the current high diversity characteristic of the tropics. While this theory is plausible and has been invoked by other authors, tests of the hypothesis have cast doubt over its wide applicability. Irrespective of the mechanistic bases for variation in species richness over wide latitudinal ranges, time is a significant determinant at a more local scale.

Birks (1980) analysed the insect faunas of trees in Britain during the period since the last glaciation 13000 years ago and concluded that there was a significant correlation with the time that a given tree species had been continuously present in Britain during that period. Taking this further, Kennedy and Southwood (1984) included Birks' radiocarbon estimates of time in Britain as a variable in a multiple regression analysis of insects on British trees.

They showed that time was a significant predictor, both in its own right and as a co-predictor with log abundance (area). This is logical considering that over time there will be a tendency for well-established tree genera to increase their abundance, despite the recent proliferation in areas arising from commercial afforestation.

SIZE

A consequence of basic structure and longevity is that trees tend to be the largest components of the phylloplane, thus providing a much wider number and range of ecological niches compared with herb and shrub layer plants. Such attributes enable more insect species to colonize the plants without facing competition, both interspecific and intraspecific.

Size has many interrelated characteristics that help to explain the presence of both herbivores and their natural enemies:

- A larger physical presence is more easily detected and colonized by invertebrates. This concept, termed 'apparency', reflects a higher visual profile for colonizing adults or passively migrating immature life stages (*e.g.* mean height was a good predictor in models of the distribution of Scolytidae in Finland; Heliovaara & Vaisanen 1995) and a greater production of olfactory cues.
- Complex plant architecture is a consequence of greater size and for trees is particularly linked to the provision of many niches capable of supporting invertebrates with a wide range of feeding strategies (guilds). Although the concept of the guild is not always clearly defined, it is nevertheless a useful means of describing the trophic interactions of invertebrates and trees without having to consider their precise taxonomic status. Thus, seed, bud, leaf, bark and wood feeders can be distinguished, each occupying a different part of the tree and therefore potentially avoiding competition.

- Increased biomass provides greater food resources per individual, thus encouraging growth and breeding success whilst reducing potential intraspecific competition. Price (1992) has reviewed the role of plant resources in insect population dynamics and concludes that, in part, the carrying capacity of an ecosystem depends on plant succession in time and space. Carrying capacity is therefore greater for trees, which generally appear late in succession, compared with herbs or shrubs. Thus biomass is available for longer and in greater quantities on trees compared with other plants.
- The concept of enemy-free space applies to the likelihood of herbivores being overlooked by searching parasitoids or predators. One aspect of size is that there is more opportunity for herbivores to occupy a part of the plant that offers protection against natural enemies (and competition from other herbivores).
- The complexity and number of niches present on a plant (plant architecture; Lawton 1983) therefore has an effect on enemy-free space. The greater the architectural complexity, the greater the time that natural enemies may have to spend in searching for prey, thus reducing their overall effectiveness and enhancing the survival probability of herbivores.

AREA

Forests tend to occupy greater areas than other plants. There is evidence to support the view that the greater the area occupied by the host plant, the greater the number of insect species associated with that host. This is called the species-area relationship and was originally linked to the island biogeography theories of species isolation and colonization.

Species-area relationships have been demonstrated for many plants and animals, including both highly mobile and sedentary species. With regard to the numbers of insect species on trees, data on the British insect fauna, one of the best characterized in the world, have been analysed by a number of authors.

Arising from earlier tentative analyses, Kennedy and Southwood (1984) analysed the relationships between insect species and characteristics of tree species in Britain. They concluded that there was a significant relationship between the area occupied by British tree species and the numbers of insect species associated with them. The area occupancy data were based on county and subcounty information of plant records and were thus potentially overestimates of actual areas occupied.

Claridge and Evans (1990) reanalysed the species-area relationships using the same insect data as Kennedy and Southwood but employing actual area information derived from Forestry Commission census data. They concluded that area was not a good descriptor for insect species on deciduous trees but was just significant for conifers.

However, further unpublished analysis of the data broken down into numbers and areas of habitat patches occupied by the tree genera indicated that whereas total area was not a good descriptor, the degree of fragmentation of that area provided a better explanation for the observed insect species numbers. A similar conclusion, concentrating on Lepidoptera that have moved hosts to include conifers, was reached by Fraser and Lawton (1994) in an analysis of new associations on British trees. The overall finding that area provides a broad descriptor of insect diversity remains valid, although this does not explain the mechanisms that lead to these associations.

TAXONOMIC RELATEDNESS

In a given ecosystem, the closer the taxonomic relatedness of host plants, the more similar the numbers of insect species associated with those plants. This has been documented for tree species that have a long history in a particular location and thus provide rich sources of invertebrate species that may subsequently colonize other tree species, especially if they are taxonomically related. A particularly good example of the process in action is the colonization of *Nothofagus,* a member of the Fagaceae newly introduced to Britain, which has been colonized rapidly by insects associated with closely related members of the native Fagaceae. Geographical isolation as a critical factor in determining invertebrate diversity and implications for international movement of pest organisms. Geographic isolation drives the acceleration of speciation and provides the means to derive principles on the processes involved. In addition, the fact of isolation means that some invertebrates abundant in one region will be absent elsewhere, despite conditions being suitable for their establishment.

In such situations, it is pertinent to examine the international plant health implications of geographical isolation and how modern patterns of trade influence the degree of isolation and later colonization by new insect species.

GEOGRAPHICAL ISOLATION AS A FACTOR IN DETERMINING INSECT DIVERSITY

In general, the basic principles apply across ecosystem boundaries and geographical zones extending from the poles to the tropics. In addition, the isolation afforded by relatively small islands provides further evidence of the principles of species colonization over space and time. Great Britain provides a particularly good example, because of its island position in the Atlantic Offshore climatic zone and the fact that it has been subjected to glaciation in recent geological time. Among the ecological characteristics that arise from this set of circumstances are a relatively depauperate mix of both animal and plant species that have arisen from the combination of losses resulting from the last ice age and the slow rate of recolonization arising from island status. British flora and fauna are therefore impoverished relative to continental Europe. Further influences arise from the preponderance of exotic tree species in commercial forestry in Britain, which although dominated by exotic conifer species also has increasing numbers of exotic broadleaved species. Expanding on the principle that colonization of tree species is determined by the features, the ranges of numbers of insect species associated with trees in Britain reflect both the area (with provisos) and length of time that trees have been continuously present in Britain. However, there are exceptions, so that Sitka spruce in particular would appear to support far fewer species than predicted on the basis of area, although this probably also reflects the relatively short history of the species and recent expansions in areas planted in Britain.

IMPLICATIONS FOR INTERNATIONAL MOVEMENT OF PEST INSECTS

Trading patterns between countries have had a considerable influence on both the flora and fauna of most of the countries in the world. This reflects deliberate introductions of plants and animals for agricultural, forestry and amenity purposes and also accidental introductions by association with other commodities. In the majority of cases, the introduced species have had a neutral to beneficial effect, reflecting their 'domesticity' within managed crop or animal husbandry systems.

However, there are many exceptions to this generalization and in most cases the results of introduction of new species were not foreseen, either through lack of knowledge of the potential impacts or because the introduction was accidental and undetected. The literature on international biological control is dominated by examples where attempts have been made to rectify the effects of new pest introductions (both animal and plant).

A classic example is the control of the woodwasp *Sirex noctilio* (Hymenoptera: Siricidae) in exotic pine forests in Australia with the introduced nematode *Deladenus siricidicola* (Nematoda: Neo-tylenchidae) isolated from *Sirex* in New Zealand. This nematode has a life cycle that includes a free-living form that feeds on the fungus *Amylostereum areolatum* and a parasitic form that invades *S. noctilio* larvae and subsequently migrates to the reproductive organs of the female wasp, causing sterility. In this situation, all elements of the interaction are exotic to the new location, with little prospect of natural control being realized.

6

Isolation and Isloation Mechanisms

Species hybridization is a very interesting subject. The successful hybridizer who produces a previously unknown combination feels that he has contributed something new, which in all likelihood would not have been produced in nature. And there is a very good chance that this new combination is also useful. Many of our most important crop plants are of hybrid origin, and new hybrid trees have definite promise.

A complete hybridization study for a single genus is a lifetime undertaking. Work by taxonomists on rare natural crosses or on introgression will provide clues as to which crosses can be made; for example, natural hybrids indicate the crossability pattern in the genus *Eucalyptus*. These clues should be supplemented by a good deal of controlled pollination work to develop a satisfactory crossability pattern because many of the most easily crossed species do not occur together in nature.

There is still much work to be done after the crossability studies, which will show that species A and B can be crossed and perhaps that the hybrids are promising. How can the hybrids be mass-produced, and which particular races or individual trees of the parent species should be used to give the best results? Those are questions the answers to which may need many years of research.

The role of hybridization varies among regions and among genera. There are groups such as *Castanea* in which the best timber species are deficient in a single character (in this case, disease resistance), which can best be obtained by crossing with another species. The majority of hybrids are intermediate between their parents in morphology and site requirements. Thus hybrids may be easily made but be of little value to forestry when large areas of intermediate habitats are not available.

Sometimes disturbance by man has created the intermediate habitats on which intermediate trees are best adapted, showing hybrid vigor, and there are examples of this in *Eucalyptus* and *Aesculus*. Again, a species sometimes finds its largest use outside its natural range, in which case hybrids may prove better adapted, an example being the hybrid *Larix decidua* x *leptolepis*. Hybridization offers relatively little hope of improving the American *Pinus resinosa*, which has resisted nearly all attempts to cross it with other species. Id the southeastern part of the United States the Southern pines are unlikely to be improved by interspecific hybridization; they reproduce readily and grow fast, while there are no other species of *Pinus* which surpass them in wood specific gravity - one of their most important properties. In any case, hybrids must be tested empirically before their potentials can be realized.

NATURAL HYBRIDS AND INTROGRESSION

Reports of natural hybrids between species belonging to the same genus are frequent. Also, there is one well-authenticated report of hybrids between tree genera, namely the cross *Cupressus* x *Chamaecyparis*. Evidence on the presumed *Tsuga* x *Picea* hybrid is less convincing.

The probability of obtaining natural hybrids is great when one sows seed collected from isolated specimens growing in an arboretum or park. If a species does not ordinarily self-pollinate under such conditions, most of the seed produced may be from crossing with another species. Also, the nursery practice is usually intensive enough to permit recognition of unusual types. Most *Abies* and *Acer* natural hybrids have occurred under such circumstances. So have a large number of those reported in *Picea, Pinus Populus* and *Quercus*.

In the majority of cases, natural hybridization leads to the production of one or a few F_1 trees only. The group of hybrids is in existence for one generation, then disappears. The next occurrence of the same combination may be in a quite different locality. This is true for most American reports of *Quercus* hybrids. It is also true of some interesting pine hybrids, such as *Pinus sylvestris* x *nigra* and *P. palustris* x *taeda*.

In *Populus, Pinus* and *Picea* study of the once-occurring natural hybrids provided the initial impetus for the later intensive work on artificial hybridization, but it was the study of the thousands of artificial hybrids which provided most information. On the other hand, study of the once-occurring hybrids of *Quercus* and *Aesculus* has contributed most of the knowledge about hybridization in those genera because artificial hybrids would be difficult to produce in quantity.

Natural hybrid swarms are infrequent but instructive. They occur most commonly on "hybrid" habitats, especially on disturbed areas. The oaks of southwestern United States furnish some of the best-studied examples (Muller, 1952). In central and northern *Texas Quercus havardii* and *Q. stellata* typically occupy sandy and clay soils respectively. Hybrids are common at the zone of contact between the two soils but not away from the zone of contact. The "pure" form of the species is found on the "pure" habitats typical of that species.

In the same region *Q. mohriana* and *Q. havardii* occur on soils derived from limestone and sand respectively. Hybrids are found where the soil is derived from a mixture of limestone and sand but not on the "pure" habitats.

In California the hybrid species *Quercus* x *alvordiana* is really a group of hybrid swarms between *Q. douglasii* and *Q. turbinella* and is found only on intermediate habitats where the two parents come in close contact. Some of these *Quercus* x *alvordiana* swarms have persisted for many generations since the close of the Pleistocene Period.

In Australia *Eucalyptus dalrympleana* was one of the components of the original forest on undisturbed sites. As a result of man's activities, many of those sites have been damaged badly by fire. Now such areas are being invaded by hybrid swarms of *E. pauciflora* x *dives* (Pryor, 1957) whereas under natural conditions the two species excluded each other because of competition. Several other examples of hybrid swarms of *Eucalyptus* have recently been found on disturbed habitats.

Aesculus sect. *Pavia* includes four species native to the central and southeastern United States. In the original forest the species remained distinct. If hybrids were formed they occurred singly or in very small groups. Now there are many road cuts and cut-over areas on which the hybrids outgrow the parents. On such intermediate habitats there are occasional hybrid swarms of *A. glabra* x *pavia, A. glabra* x *octandra,* and *A. octandra* x *sylvatica.* In the latter two cases hybridization has led to considerable gene flow from *A. octandra* to the other two species. In yet another combination, *A. pavia* x *sylvatica* there is considerable gene flow in both directions from one parental species to the other.

By introgression is meant an extended period of hybridization followed by a considerable flow of genes from one species to another. Frequently it is hard to distinguish between the coming together of species by introgression and the divergence of species or races. In *Picea* introgression is suspected of being a majoı factor contributing to the wide adaptability of the northern, circumpolar species although it was proved in only two cases (Wright, 1955); *Picea engelmannii* and *P. glauca* cross readily in interior British Columbia and Alberta. Evidence of one species is found in the other far from the zone of contact.

The spruce population of northern Scandinavia and adjacent parts of the U.S.S.R. is a mixture of intergrades between *P. abies* and *P. obovata*. In other cases widely distributed species with adjacent ranges are known to cross artificially and are presumed to have crossed naturally when climate favored a range overlap. On the other hand, lack of genetic variability and lack of recent opportunity for hybridization are characteristic of the widely scattered southern *Picea* species.

Following the introduction of Populus deltoides from North America into southern Europe there has been a large amount of introgression from that species into *P. nigra*. Pure *P. nigra* is now rare in many localities and planting stock consists mostly of hybrids (Schreiner, 1959).

True introgression is rare in *Quercus* despite the large number of species which have been reported. Intermediates between the north European species

Quercus robur and *Q. petraea* occur where their ranges overlap and may be hybrids. In southwestern United States *Quercus* x *drummondii* seems to be an introgressive derivative of *Q. stellata* with an infusion of *Q. gambelii* genes. In most other cases the species have maintained their identities even when growing in mixture with several other oak species (Palmer, 1948).

Three factors help to explain why hybridization sometimes leads to only a few intermediates and sometimes to hybrid swarms or introgression. First, there may be limited opportunity for crossing.

A few natural hybrids have been reported from arboreta for the cross between *P. strobus* from eastern America and *P. griffithii* from the Himalayas. These hybrids frequently appear in northern Italy where the two species are often grown together in gardens and parks. The cross is easily made artificially and introgression might be expected if the parents had the same range. Differences in flowering time may be all that keep *Acer rubrum* and *A. saccharinum* from becoming one species. Natural hybrids have not been reported although the two species commonly grow together and can be crossed with ease. Secondly, the hybrids may be sterile. As yet there are few instances in which this is known to be the case.

The third factor is the nature and amount of selection pressure. Suppose that genes A and a are allelomorphic to each other, with respective frequencies of p = .9 and q = .1. Under random mating the zygotes will be present with relative frequencies of $(p + q)^2 = p^2 + 2pq + q^2 = .81AA + .18Aa + .01aa$. If selection favors the heterozygotes for one generation, the proportion of *Aa* trees will increase to a maximum (the theoretical maximum being .50).

But if environmental conditions then change so that selection favors the homozygotes, the Aa trees will tend to disappear. With complete elimination of heterozygotes the hybrid group in which zygote frequencies were *.81AA: .18Aa: .01aa* will produce gametes (pollen and egg cells) in which the gene frequency is *.988A: .012a.* These gametes will produce a new generation of trees in which the zygote frequencies are *.9757AA: .0242Aa: .0001aa.* Notice that selection against the heterozygotes has reduced the frequency of *Aa* trees by about 90 percent and reduced the frequency of the less common homozygote (*aa*) by about 99.9 percent. In other words, very few a genes remain with which to form *Aa* trees in the following generation. If the direction of selection should again be reversed to favor *Aa* trees, the build up of a hybrid swarm would have to start afresh.

This mathematical example has two corollaries. First, the existence of a persistent hybrid swarm is strong evidence that selection favors heterozygotes or, in other words, that usable hybrid vigor is present. Second, the absence of a hybrid swarm does not necessarily mean that useful hybrids cannot be produced. Man can take advantage of hybrid vigor exhibited in two generations out of three whereas nature cannot.

If a breeder's interest is in the overdominance indicated by the presence of a hybrid swarm, he should attempt to mass-produce by artificial means F_1 or possibly F_2 populations because it is in those generations that the overdominance will be most pronounced. However, if he is not in a position to undertake such an intensive program he should not attempt an intermediate measure such as a genetic thinning. By another theorem from population genetics, it can be shown that such a hybrid population will come to a natural equilibrium when the

$$\text{frequency of the a gene} = q = \frac{s1}{s1 + s2}$$

s1 and s2 being the selection pressure against the *AA* and *aa* trees respectively. It can also be shown that the population is most productive at this natural equilibrium (if it cannot be maintained in a completely heterozygous state by artificially crossing *AA* and aa trees). As an example, if *AA*, *Aa*, and aa trees have fitness ratings of .60, 1.00, and .85 respectively, the hybrid population will have an average fitness of .89 if at equilibrium. Artificial selection to favor the best trees (which would be the heterozygotes) would increase the number of unproductive *AA* trees and would lower the average fitness to .862 whereas artificial pollination to ensure that only *aa* trees were planted would increase the average fitness to 1.00.

If a breeder's interest is primarily in additive genetic variance, a natural hybrid swarm forms an excellent starting point for a selection program. In southern Michigan there is a large group of hybrids between *Pinus thunbergii* and *P. densiflora.* In several characters the additive genetic variance within it is probably two or three times that found in any natural population of the pure species. Heritability and response to selection would also be two or three times as great as in a population of the pure species.

CROSSABILITY PATTERNS

With the exception of *Cupressus* x *Chamaecyparis*, mentioned earlier in the chapter, intergeneric hybrids in temperate zone trees can be disregarded and attention can be focused on crosses within genera. The crossability patterns are best known in *Populus, Pinus Picea Quercus, Castanea,* and *Larix*. In those groups enough crosses have been attempted to formulate general rules which permit one to forecast which species combinations are most likely to be successful. For *Acer, Aesculus, Betula, Fraxinus* and *Ulmus*, information is more fragmentary and permits only the broadest generalizations. For most other genera (including all the tropical ones) it is even more fragmentary, consisting mostly of lists of natural hybrids.

The amount of genetic differentiation within what is considered to be a single genus varies greatly. At one extreme is *Populus*. There has been so little differentiation that a given species can apparently be crossed with any other species. *Castanea* may have a similar pattern - all attempted species crosses have been successful. Genetic differentiation is a little more pronounced in *Picea*. Most species can be crossed with their geographic neighbors but not with those species from a distant locality.

There is no sharp break into distinct crossability groups and it is possible (but this has not been tested) that any species could be linked to any other one by crossing with intermediate types. *Eucalyptus* exhibits a similar pattern, neighboring species in the same section crossing with relative ease. *Quercus* is divided into 3 subgenera (only 2 are known genetically). A large number of hybrids between species belonging to the white oak subgenus *Lepidobalanus* have been reported and possibly any two white oaks could be hybridized. The same is true of the red oak subgenus *Erythrobalanus.* On the other hand, there are no convincing demonstrations of hybridization between the subgenera of *Quercus.*

The pines are at the other extreme. The genus *Pinus* is divided into 2 subgenera and a large number of series. With 3 exceptions attempts to cross members of the same series have been successful. The pattern in *Acer* is similar to that in the pines. Of 60 artificial crosses which have been attempted, 6 succeeded.

The successes were all within one of the 13 recognized sections. From the genetic standpoint, series or sections in these 2 genera have approximately the same amount of differentiation as some entire genera. It would make sense biologically (but not practically) to regard these small groups of species as separate genera, as some authors have done in the past.

Morphological or physiological similarity and geographic distribution provide excellent clues as to the crossability pattern in a genus. The most easily made combinations are between similar species with nearly overlapping ranges. Often distance was a sufficient barrier to render unnecessary another type of isolating mechanism. If this barrier is removed by interplanting or by controlled pollination, hybrids result. On the other hand, species which occur in mixture in the same stands usually do not cross.

Sometimes differences in flowering time or in microsite requirements provide an isolating barrier which can easily be overcome by artificial crossing. Sometimes the isolating barrier arises as the result of genic or chromosomal differentiation. Irradiation of pollen and the use of pollen mixtures offer some promise of overcoming such a barrier. It is also difficult to cross morphologically dissimilar species or species with widely separated distribution areas. In such cases, long separation has usually been accompanied by great genie or chromosomal differentiation.

Crosses which fail in one environment may succeed at another place or time. Knowledge of the nature of the barrier often helps to break it down.

Examples of the relationship between crossability, similarity, and geographic distribution are numerous, and for illustration the genus *Pinus* can be cited. The series *Strobi* comprises several similar species. All except one have separate (but neighboring, if considered in the geological sense) distribution areas and can be crossed with each other rather easily. The single exception (*Pinus lambertiana*) is not typical of the series and does not cross with the other members. Several attempts have been made to cross species of ser. *Strobi* with members of other series but only one such attempt succeeded.

Pinus ser. *Sylvestres* comprises about a dozen hard pines. The Asiatic members are morphologically similar and can be crossed with each other in nearly all possible combinations. They are less similar to the European representatives of the series. Only a few European x Asiatic crosses successful and these gave low seed sets. The single United States representative in series *Sylvestres (P. resinosa)* has several distinctive morphological and chromosomal characters. Over 20 years of effort by many people resulted in only 4 hybrid trees between *P. resinosa* and another species.

Pinus ser. *Australes* comprises groups of southeastern and southwestern United States species. Members of each group are more similar to each other than to members of the other group. All possible southeastern x southeastern species combinations and a great many southwestern x southwestern combinations have resulted in hybrids. However, no southwestern x southeastern combination has been successful (Duffield, 1952).

The spruces (*Picea*) also show a relation between crossability, similarity, and distribution. In northern United States, Canada, and northeast Asia is a group of 6 species, each with a nearly separate distribution area and each with certain similarities to its neighbors. All attempted hybridizations within the group were successful but several attempts to cross those spruces with others outside it were not. Across northern Europe, Siberia, and northern China is another "chain" of similar species, within which crossing occurs very easily. The 6 species which inhabit the same mountain in Japan do not cross, but one of them can be crossed with mainland spruces. One of the notable exceptions to the general crossability pattern is the cross *P. omorika x sitchensis*, which has been successful in Europe.

The genus *Larix* contains about 10 species, each with its own separate geographic range. Apparently differentiation has been entirely geographic and the species cross easily.

Of 7 attempted artificial hybridizations in *Fraxinus* only one succeeded. That was between an eastern United States and a western United States representative of the same section. The 2 species can be considered neighbors in the geological sense and are so similar that one could mistake aberrant specimens of one for the other. Hybrid swarms are reported between members of a similar species-pair in southwestern Europe.

Several cases are known in which crosses between species with different chromosome numbers yielded good seed. The general relationship between chromosome number and crossability has not yet been worked out.

These generalizations can be summarized by pointing out that a good monograph and a detailed set of range maps are of great help in planning a species crossing program. If it is assumed that there are 90 species in the genus *Pinus*, there would be a total of (90 x 89)/2 = 4,005 different species x species combinations. Work to date indicates that 10 percent or 400 of these might be successful. A breeder who crossed at random would have only a 10 percent probability of success. But one who studied the taxonomy and distribution could increase this probability to 80 percent.

BREEDING FOR HYBRID VIGOR IN THE F1 AND F2 GENERATIONS

Most emphasis in species hybridization has been on the production of vigorous F_1 hybrids. Enough work has been done to show that the approach is extremely promising. The majority of the hybrids are useless but enough possess hybrid vigor to make the work worthwhile. Often hybrid vigor is evident in one part but not in another part of the range of a species.

The introduction of a successful F_1 hybrid combination into commercial practice is carried out in four stages. First comes a crossability study to determine what combinations can be made. As already indicated, this work is well advanced in some genera and scarcely started in others. Next comes the production of a few hybrids of each promising combination and their testing in comparison with the nonhybrids they might replace. A good start has been made in this direction, but in scarcely any case have the tests been extensive enough to indicate a hybrid's true potentialities. Most of the *Pinus* hybrids produced in California and Pennsylvania are scarcely known outside a few test localities; and the potentialities of *Larix decidua x leptolepis*, which is very useful in Europe, are scarcely known in America. The third stage is the determination of mass production possibilities.

This is no problem in those species of *Populus* which can be propagated readily by cuttings, however, it is a very real problem in other genera. The Institute of Forest Genetics at Placerville, California, did enough work on several combinations of *Pinus* to show that mass production is possible. In recent years, the Korean Institute of Forest Genetics has done a large amount of work, and obtained extremely promising results, on a few *Pinus* combinations. The fourth step in introducing a successful F_1 hybrid into commercial practice is determining which particular trees would make the best parents.

This has barely started for any species combination. Take the cross *Pinus sylvestris x densiflora* for example. It was produced in the United States, originally in Connecticut and later in Pennsylvania. In both cases the first available trees, of unknown origin, were used as parents. The hybrids grew extremely fast but cannot be used because of poor form and low seed set.

The Pacific southwest and the northeastern forest experiment stations of the United States Forest Service have produced a large number of species combinations in *Pinus* but only a few can be mentioned here. *Pinus monticola x strobus* is one of the most promising. Seed sets are high enough to indicate good mass production possibilities; growth in California and in many other parts of the United States equals or exceeds that of the faster-growing parent; there is intensive work under way on breeding disease-resistant cultivars of both parent species. When the work is completed, truly remarkable F_1 hybrids will be possible. The hybrid *P. attenuata x radiata* also shows promise. It could be mass-produced. Although it does not outgrow the faster-growing parent *(P. radiata)* on warm sites, it promises to be one of the most productive pines on sites too cold for that species.

Seed sets are very high for the hybrid *P. densiflora x thunbergii*. This combination shows considerable hybrid vigor in Pennsylvania, but not in Japan where the parents are native. The potential usefulness of a fourth combination, *P. nigra x resinosa*, is limited because of mass production difficulties. Only four hybrids were produced after many years' effort. Thus, these hybrids have good growth characters but remain botanical curiosities.

The potentialities of *F. pine* hybrids are being exploited to the full in Korea. There thousands of bags are used to make the cross *P. rigida x taeda* each year and the program is proving to be economic. The hybrids can produce the most income even though they are more expensive to produce than nonhybrids.

The cross *Larix decidua x leptolepis* is the subject of more than 27 different publications from as many parts of northern Europe, almost all reporting greater growth than for either parent species (for the references see Wright, 1962). Seed sets are high but *Larix* does not lend itself as well to mass-controlled pollinations as does *Pinus* Syrach Larsen of Denmark has therefore designed a seed orchard containing a self-incompatible clone of one species and seedlings or clones of the other. The self-incompatible clone serves as the female parent and produces hybrid seed.

A small amount of American and a much greater amount of Russian work has resulted in the production of several artificial hybrid combinations in *Quercus*. Some of these are intermediate between their parents but others exhibit useful hybrid vigor (Piatnitsky, 1960). However, at present even the latter appear to be of very limited usefulness. Controlled pollination is extremely tedious and likely to be too expensive to be done on a commercial scale. Vegetative propagation is impractical. Seed orchards where natural crossing can take place may be the answer, but efforts to find trees which set large percentages of hybrid seed after natural gross pollination have been unrewarding.

The silviculture of *Populus* is now based much more on hybrids than on pure species. Cultivars derived from crosses involving *P. deltoides, P. nigra P. trichocarpa* and other species are commonly planted in Europe. Most of these have not been compared with their parents but hybrid vigor is usually assumed. As the result of Schreiner's pioneer work in the northeastern United States, many promising hybrids have been tested in America and can soon be planted on a commercial scale.

Species of *Populus* other than the aspens can be propagated readily by cuttings so that mass production is no problem. Ease of rooting has also permitted breeders to conduct a large number of tests for clonal differences within F_1 full-sib families. As expected, significant clonal differences were rarely found except when the parents were themselves hybrids.

In short-term tests *Acer platanoides x mayrii* and *A. platanoides x cappadocicum* exhibited some hybrid vigor. The crosses were easily made on an experimental basis at the rate of a few hundred per day. However, that rate could not be increased greatly because of the flower structure, and mass-controlled pollination does not seem feasible. Neither do seed orchards because of differences in flowering time. Polyploidy is possibly the answer. Perhaps the few hybrids now available could be made truebreeding by having their chromosome number doubled. But would the tetraploids be as desirable as the diploids? That must be determined.

Theoretically one can expect a certain amount of hybrid vigor to persist into the F_2 generation. This raises the possibility of utilizing the F_2 generation if the F_1 cannot be produced cheaply. At least in some of the pines the F_1's are sufficiently fertile for F_2 seed to be produced in quantity. The Institute of Forest Genetics in California has been pursuing this possibility for several Pinus, combinations. In the United Kingdom there are plans to raise large F_2 progenies of European x Japanese larch *(Larix decidua* x *leptolepis)* in the expectation that some hybrid vigor will persist.

In most of the illustrations quoted here, the hybrids are still young, since there are no adequate older tests of many of the most interesting combinations. Also, many of the preliminary growth tests have been conducted on only one or two sites, and it is well known that hybrid vigor may be evident at one place and not at another. There is, therefore, considerable testing still to be done.

Practically all the *Pinus, Picea* and *Populus* hybrids to date have been made with the first available parents. This leaves considerable work for future tree breeders. The first *Pinus, densiflora x thunbergii* hybrids produced in Philadelphia seem quite useful but it is inconceivable that they can be the best. It is much more likely that even better hybrids can be produced by crossing selected races of the two parent species or perhaps selected individuals within those selected races.

POSSIBILITIES OF RECOMBINATION IN LATER GENERATIONS

Almost every tree hybridizer hopes he will produce an F_1 or F_2 generation which possesses enough hybrid vigor to be worth planting on a large scale. That is not always possible. Sometimes segregates arising in a later generation will be the valuable product. This will be true under three conditions:

1. If controlled pollination is so difficult that F_1's and F_2's cannot be mass-produced, (as in *Juglans, Castanea, Acer, Quercus* and possibly *Abies*);
2. If hybrid vigor is present but is not sufficient to make a plantable cultivar *(Castanea dentata x mollissima);*
3. If the genetic variance is mostly additive.

If it is desired to incorporate one character from species A with many characters from species B, back-crossing is indicated. The F_1 hybrid between A and B should be produced artificially, then backcrossed to species B and the process repeated again if necessary. In crop plants four or five generations of backcrossing are frequently necessary. If it is desired to incorporate approximately equal amounts of species A and B, an F_1 generation is produced; the F_1's are crossed with each other to produce an F_2, those are crossed with each other to make an F_3, etc.

The necessity for this approach has been appreciated for many years but investigators have been loath to follow it because of the long periods of time involved. Such delays may be necessary but need not be interminable. Numerous, important tree species produce flowers at ages from one to five years. If a tree breeder had known their desirability at the start, within 20 years he could have had F_4 generations in production.

At present most many-generation breeding projects are unplanned developments of F_1 hybridization studies. This situation must be changed if the most rapid progress is to be made. The field tests designed to produce a later generation segregate should be different from those designed to test the usefulness of the F_1 or F_2 generation *per se*.

The aim in a many-generation project should be to produce and test an F_3 or second generation backcross as rapidly as possible. Why? The F_1 generation is a uniform one, offering almost no opportunity for selection. [2] Suppose that there are 25 F_1 seedlings and 10 of them fruit at a very early age. The genetic component of the variation in the F_1 population is almost nil, so that it makes little difference whether those 10 are used as parents of an F_2 population or there is a wait of several years in order to see whether the other 15 might have superior phenotypes.

The F_2 population is a genetically variable one and offers opportunity for genetic selection. But a large proportion of the F_2 genetic variance is nonadditive and therefore cannot be fixed by crossing the best trees. This means that a breeder is justified in getting only a moderately good evaluation of the F_2 because no matter how long he looks at the trees he cannot tell their breeding potential as well as if he were to wait until the F_3 or F_4.[3] It is better to look at the F_2 for relatively few years and to practice a mild selection. The 10 or 20 percent best trees might be chosen and used as parents of a replicated F_3 test in which the identity of all families is maintained.

It has already been mentioned that selection is more productive in the F_3 than in the F_2 generation because of the smaller amount of nonadditive genetic variance. Another reason for preferring the F_3 is that it offers an opportunity to practice family selection. This means selecting breeding stock on the basis of a tree's phenotype as well as on the performance of other trees having the same parent(s). This is much more effective than selection on the basis of phenotype alone.

Should one use control-pollinated (both parents known) or open-pollinated (only the seed parent known) progenies when conducting a selective breeding program such as outlined here? The F_1 and F_2 populations can be relatively small, consisting of a few dozen or few hundred trees. They could well be control-pollinated. As long as a tree breeder is working with multiple factor inheritance (which is most likely), the F_3 population should be very large, containing scores or hundreds of progenies and thousands of individual seedlings.

The likelihood of success with small F_3 tests is too small to warrant the trouble involved. In *Populus Picea*, and many *Pinus* species controlled pollination can be easily done on the required scale. On the other hand, workers with genera in which controlled pollination is laborious, as in *Abies* in which the female flowers are rather inaccessible, or *Castanea Juglans* or *Acer* in which each pollination yields few seeds, would be well advised to work with open-pollinated progenies in order to have tests of adequate size.

Vegetative propagation is commercially possible with many species of *Populus*. Hence, in that genus it may be desirable to establish clonal tests in the F_2 or F_3 generations in order to select clones with desirable properties. In genera such as *Pinus* and *Quercus*, clonal tests are expensive and probably not needed because seedling families will be planted commercially.

Family selection in segregating generations will play a large role in the *Castanea* breeding program. The American *C. dentata* is a fast-growing timber tree with desirable growth habit and wood properties but without resistance to chestnut blight *(Endothia parasitica)*. That must be furnished by the Asiatic *C. mollissima* or *C. crenata*, neither of which becomes a large timber tree. A moderate number of F_1 hybrids have been produced which show more or less intermediacy between the parents. Also, moderate numbers of F_2's and backcrosses have been produced. Some of the F_2 segregates have good vigor and much more resistance than the *C. dentata* parent. That is not enough, however, and it appears that large F_3 and F_4 generations must be raised to yield a cultivar which is commercially acceptable for timber production.

A great deal of the *Populus* breeding work in southern Europe is probably aimed toward the production of desirable later-generation segregates. However, no formal many-generation program has been started, selection being among trees of unknown generation.

Several F_1 *Pinus* hybrids have reached fruiting size at the Institute of Forest Genetics in California. Most have been fertile. One of the most promising crosses is *P. jeffreyi* x *coulteri.* Some F_2's and backcrosses exhibit the desirable growth characters of the former and resistance to the pine reproduction weevil *(Cylindrocopturus eatoni)* of the latter. The degree of recombination is sufficient to indicate that a group of F_3 families might comprise a commercially acceptable cultivar.

INTERRACIAL HYBRIDIZATION

The crossing of different races within a species offers the same possibilities as does the crossing of different cultivars of a crop plant or different breeds of animals. The F_1 hybrids may exhibit hybrid vigor as do crosses between inbred lines of maize. Or they may exhibit a combination of desirable characters and furnish the foundation for a several-generation selection program.

The pioneer German tree breeder, Dengler, made a start on interracial hybridization in 1926 when he crossed German, French, and Scottish origins of *Pinus sylvestris* He thought some combinations showed hybrid vigor at 10 years old, but Scamoni later reported, when the trees were 20 years old, that the hybrids were intermediate between their parents.

In 1948 Johnsson and his co-workers at Ekebo, Sweden, made a systematic series of crosses among European aspens *(Populus tremula)* originating at three different latitudes in Sweden. At 10 years all the hybrids were considerably superior to the midparent mean and one combination (southern x central) was taller than the faster-growing parent.

Other Swedish tree breeders, especially Nilsson and Langlet, have had promising results from interracial hybridization in *Pinus sylvestris* and *Picea abies*. These were seen during the study tours preceding the world consultation. Following their lead, workers in other regions will undoubtedly show greatly increased interest in this subject.

FACILITIES NEEDED FOR HYBRIDIZATION WORK

Good collections of breeding material are the prime requisite for productive hybridization work. The breeder interested in determining a crossability pattern needs several flowering specimens of several races or species. The breeder interested in mass-producing a particularly promising hybrid needs many selected specimens of each parent. One or two specimens of each parental type are not enough because they may not fruit nor give enough nonhybrid seed to provide controls.

In temperate climates a period in the spring should be reserved for at least five years when embarking on a serious hybridization program. Time is needed for bagging, pollinating, removal of bags and observation of flower development. Seed sets are often so variable that the same cross needs to be repeated a few times to establish a reliable average pattern.

A hybridizer needs a small but good research nursery and several test sites. Usually few seeds are available of combinations which are made for the first time. This means that the testing must be done with greater precision than is usual in planting research. Both nursery and experimental forests must, therefore, be equipped to give the tests better than average care.

If the breeder is interested in developing both F_1 and F_3 cultivars, duplicate test plantings may be necessary. The design used to determine comparative growth rates of the F_1's and their parents may not be the one which would insure earliest fruit production. Also, a good design to test hybrid vigor would not necessarily be the one used to obtain F_2 and F_3 hybrids by natural open pollination.

Bibliography

Aukland, L.: *The Effect of Selective Logging on Amphibian Diversity of Budongo Forest*, Uganda. B. A., Oxford University, 1997.

Bahati, J.: *Impact of Arboricidal Treatments on the Natural Regeneration of Species Composition in Budongo Forest*, Uganda. M. Sc., Makerere University, 1995.

Barrows, M.: *A Survey of the Intestinal Parasites of the Primates in Budongo Forest*, Uganda. Glasgow University, 1996.

Chakrabarty, Falguni: *Adaptation of the Santals to the Hill-Forest Environment*, APH, Delhi, 2012.

Fairgrieve, C.: *The Comparative Ecology of Blue Monkeys* (Cercopithecus mitis stuhlmanni) in Logged and Unlogged Forest, *Budongo Forest Reserve*, Uganda, Edinburgh University, 1995.

Jerram. M.R.K. : *A Text-Book on Forest Management : In Agriculture, Horticulture and Forestry*, Asiatic Pub, Delhi, 2006.

Johnson, K.: *Hunting in the Budongo Forest*, Uganda. SWARA, 1996.

Johnson, K.: *Local Uses of Budongo's Forest Products*. M. Sc., Oxford University, 1993.

Miller, P. L.: *Some Dragonflies of the Budongo Forest*, Western Uganda. Opusc, 1993.

Musamali, P.: *The Ecology, Diversity, and Relative Abundance of* Small Mammals in Primary and Disturbed Forest Compartments *in the Budongo Forest Reserve*. M. Sc., Makerere University, 1996.

Owiunji, I.: *The Long Term Effects of Forest Management on the Bird Community of Budongo Forest Reserve*, Uganda. M. Sc., Makerere University, 1996.

Reynolds, F.: *Chimpanzees of the Budongo Forest*, New York, Rinehart and Winston, 1965.

Reynolds, V.: *Budongo: A Forest and its Chimpanzees*, London, Methuen, 1965.

Sathe, T.V.: *A Textbook of Forest Entomology*, Daya, Delhi, 2009.

Sharma, Ajay: *A Text Book of Forest Trees Entomology*, InternationalBook Dist, Delhi,

2007.

Shiel, D.: *The Ecology of Long-term Change in a Ugandan Rainforest*. D. Phil., Oxford University, 1997.

Solvang, R.: *Edge Effects on an Understorey Bird Community in Budongo Forest Reserve*, Uganda, Agricultural University of Norway, 1996.

A L Bhatia : *Biochemistry and Endocrinology*, Indus Valley Publication, Delhi, 2002.

Akhilesh Sharma : *Biochemistry*, RBSA Publication, Delhi, 2006.

Arnold C. Long : *Advances in Seafood Biochemistry*, Cyber Tech Publication, Delhi, 2009.

Arun Rastogi : *An Introduction to Practical Biochemistry*, Anmol Publication, Delhi, 2010.

Arvind Shukla : *Analytical Biochemistry*, Discovery Publishing House, Delhi, 2009.

Ashok Kumar Sharma : *A Textbook of Biochemistry*, Random Publication, Delhi, 2012.

Ashok Kumar Sharma : *Animal Biochemistry*, Random Publication, Delhi, 2012.

Ashok Kumar Sharma : *Biochemistry of Biomolecules*, Random Publication, Delhi, 2011.

Bandana Ghosh and S.N. Mukhopadhyay : *A Textbook of Biochemistry*, Wisdom Press, Delhi, 2012.

Bimla Singh : *Biochemistry and Genomic Revolution*, Vista International Publication, Delhi, 2006.

Chanderlekha Goswami : *Biochemistry and Instrumentation*, Manglam Publication, Delhi, 2011.

Chiranjib Chakraborty : *Advances in Biochemistry and Biotechnology*, Daya Publication, Delhi, 2005.

G.P. Garg : *Biochemistry of Catalysts*, Discovery Publication, Delhi, 2010.

Hammond : *Plant Biotechnology: New Products and Applications*, Springer, Publisher, Delhi, 1998

Hari P Singh : *Experimental Plant Biotechnology Techniques*, Agrihortica Publication, Delhi, 2007.

Harsh Bhaskar : *Basic Facts on Biochemistry*, Campus Books, Delhi, 2010.

Index